The
UNIVERSE OF REALITY

The
UNIVERSE OF REALITY
A Guidebook to Principle

A. Edward Stinson

iUniverse, Inc.
New York   Bloomington   Shanghai

The UNIVERSE OF REALITY
A Guidebook to Principle

iUniverse books may be ordered through booksellers or by contacting:

iUniverse
1663 Liberty Drive
Bloomington, IN 47403
www.iuniverse.com
1-800-Authors (1-800-288-4677)

ISBN: 978-0-595-48881-0 (pbk)
ISBN: 978-0-595-60882-9 (ebk)

Printed in the United States of America

This writing is created for, and dedicated especially to Frances, Daniel, Ronald, Gary, Steven, Stephanie, Shannon, Lauren, and Bobby, and all my family; also, to our Scientists, Physicists, and Astronomers.

These seeds are humbly submitted
in hopes they may grow free and giant trees.

Preface

The writings in this volume of books are a combination of diverse subjects that are each too short to individually produce in print. They are alternative physics, politics, religion, and poetry. The primary subject is alternative physics, which is also the largest book. All four subjects are controversial and are the studies and *opinions* of the *author alone*. This volume contains many theories that are contradictory to the presently accepted theories. Contrast is not always good; but, it is usually interesting and fun.

Theories and conjectures are like cars: they are designed, built, tested, displayed, investigated, used, abused, repaired, beat up, banged up, and finally traded in for new ones when they don't work so well anymore. Eventually accepted theories are replaced by "better" ones. New theories are revoked, hated, and rebuked, but when tested thoroughly they may be slowly accepted.

The following writings are definitely antiestablishment.

Contents

Introduction

Infinity is the limit of dreams. Dreams create theories; egos reject them. The unified force concept in Book 1 (The Universe of Energy) will be pushing aside some long-accepted theories of Einstein, Newton, and others; however, I do not disparage, but esteem the integrity of their reasoning. In fact, this writing struggles to build on their foundations. Without Newton, Einstein, and many other dedicated Scientists, the concepts presented in this writing would not be possible. Albert Einstein would probably be delighted with these concepts, for they should complete the search for his "unified" theory; subsequently, reaffirming his belief in nature's dice. Concerning Sir Isaac Newton, this concept corrects his simple mistake in assuming "gravity's" origination, or direction.

An answer for Albert is now set before science, a puzzle with all the pieces numbered and assigned, when composed shall reveal the integral quintessence of the universe.

The concepts presented in this writing are like bathyspheres bobbing on the surface of a very deep ocean, preparing to submerge, observe, and discover a new world suspected, but never revealed.

Our literary subs must dive into dark and strange depths to understand light, dark, and the inside and outside universes. The Universe Of Energy (UOE) concepts are only vent valves allowing us to descend to higher understanding of our universe. The explanation of gravity, as given by the theory of the UOE is somewhat reversed compared to most accepted theories. The UOE holds that gravity does not originate from mass, but rather from the energy that saturates the universe. The planets, stars, galaxies, etc. simply shield part of this energy from each other causing these particles to be forced toward each other, rather than "pulled" together. Please remember that many of the presented numerical values may not be entirely accurate. Only stringent, unflawed, empirical research can determine true values. And then, all values in science are only interpretations of studies.

BOOK 1

The UNIVERSE OF ENERGY
Alternative Physics and Science

Phase I
The **DARK** Side of Light

Phase I introduces, deciphers, and untangles light wave duality phenomena and light pattern interference fringes, subsequently explaining the fact that electromagnetic energy (EME) is both light, dark light (mass), and gravity. This includes the entire EME spectrum.

Most earthlings accept the statement "$E=mc^2$." The majority agree with it, but few have attempted to explain in depth why and how mass is energy and vice versa. What does velocity, especially at the very fast rate of c^2, have to do with mass and energy exchanging characteristics? Phase I theories prepare, and glue together, the other Universe of Energy (UOE) theories by presenting logical, mathematical explanations, and descriptions of "$E=mc^2$" or transposed; "$m=E/c^2$," and how mass "absorbs" energy to allow shielding of the UOE to effect (produce) gravity.

Decimals and Review

We should have a brief discussion of decimal places and scientific notation at this point, before attempting basic physics. There are hundreds of units and symbols used throughout physics. To our dismay, it appears that many diverse areas of physics use the same symbols, but with different meanings and representations. This presents a problem when combining these separate areas of physics. Therefore, to eliminate the conflicts between these areas (and for our convenience) a few original symbols are provided, which may be found in various areas of physics, such as astrophysics, nuclear physics, and quantum electrodynamics.

The calculations presented in this writing are not exact because references from various sources have been used. All physics and science sources do not agree on masses, densities, distances, velocities, etc. of particles.

The UOE uses average dimensions and quantities, therefore a good portion of the conclusions are ballpark figures. When dealing with very large figures having from ten to fifty zeros behind them there will be a large effect when rounding off, even if the tenth or twentieth place is rounded off. For instance, if there is a number with twenty zeros (100,000,000,000,000,000,000 or $1.0E^{20}$) and it is rounded off at the seventh zero, it will affect over a trillion values.

For folks that are not familiar with scientific notation, there will be a short explanation, and you lucky folks that are familiar with it can go ahead and take a two minute break.

When we have large numbers and do not wish to write them out we can count digits and record their number. For positive numbers greater than one, such as 34,567,000,000,000.00 we put the decimal after the first number and record the number of whole numbers or zeros behind it, and we get $3.4567E^{13}$ where the "E^{13}" is plus (+) and tells us we have moved our decimal 13 places to the left. Now, if we have a large fractional number such as .00000000000034567 and we move the decimal to the position behind the first whole number, we get $3.4567E^{-13}$, where this time we've moved the decimal 13 places to the right, and we also have a negative notation (-13).

Ballpark figures can be as close to home plate from the pitcher's mound or as far out as to second base, center field, or even out in left field, depending on what reasoning they are based on. Ballpark figures are O.K. as long as there is an umpire to make the call should a figure get knocked out of the park. For brevity most figures will be kept to about four or five decimal places.

There are names and symbols for commonly used scientific notational prefixes. These are listed in Table I-1.

Table I-1 starts from the smallest number, where a decimal has been moved to the right to obtain scientific notation, and continues to the larger numbers where the decimal has been moved to the left. In scientific notation the decimal is always placed behind the first whole number.

Table I-1

prefix	symbol	E	number represented
yocto	y	$1.0E^{-24}$	.000000000000000000000001
zepto	z	$1.0E^{-21}$	.000000000000000000001
atto	a	$1.0E^{-18}$	.000000000000000001
femto	f	$1.0E^{-15}$	.000000000000001
pico	p	$1.0E^{-12}$	.000000000001
nano	n	$1.0E^{-9}$	.000000001
micro	m	$1.0E^{-6}$	.000001

prefix	symbol	E	number represented
milli	m	$1.0E^{-3}$	.001
centi	c	$1.0E^{-2}$	.01
deci	d	$1.0E^{-1}$	.1
one		$1.0E^0$	1
prefix	symbol	E	number represented
deca	da	$1.0E^1$	10
hecto	h	$1.0E^2$	100
kilo	k	$1.0E^3$	1,000
mega	M	$1.0E^6$	1,000,000
giga	G	$1.0E^9$	1,000,000,000
tera	T	$1.0E^{12}$	1,000,000,000,000
exa	E	$1.0E^{15}$	1,000,000,000,000,000
peta	P	$1.0E^{18}$	1,000,000,000,000,000,000
zetta	Z	$1.0E^{21}$	1,000,000,000,000,000,000,000
yotta	Y	$1.0E^{24}$	1,000,000,000,000,000,000,000,000

Light Speed Variations

The duality of light (wave/particle) theory and their empirical studies are very confusing and not easily understood by scientists, much less laymen. Electromagnetic energy (EME) is, and must be continuous, yet a true continuous wave cannot have a Doppler effect. Actually, the Doppler effect is created by pulses and velocity. Energy and its associated mass cannot exist simultaneously, although mass and energy are associated through velocity ($E=mc^2$). Therefore, velocity must be the connection, and the divider, of energy and mass.

Electromagnetic energy (EME) consists of, at least, sixteen (16) basic characteristics:

1. Electric field
2. Magnetic field
3. amplitude
4. wavelength
5. wave shape
6. velocity
7. polarization (polarity)
8. mass
9. spin
10. energy
11. position
12. direction
13. density
14. volume
15. charge
16. force (pressure)

Most people, professors, and laymen, are somewhat familiar with the concept of photons as "bundles" of energy. Actually, they are more like bundles of "mass." Scientists and physicists for many years have been experimenting with light and its properties. Some experiments show light as a wave function and others show it as a particle function. Light appears to be a wave and a particle because *it is both*, but not at the same time; it alternates (oscillates) between states.

Generally, light is mass a small part of the time, and energy the rest of the time. The energy part of a "wave packet" is only energy due to its greater than light velocity (*c*). The mass part is only mass for its lack of velocity, as energy becomes a particle at sub-light velocity. This particle will increase in mass as its velocity is maintained below c, decreasing to zero velocity where the maximum mass occurs, at zero velocity. This velocity variation occurs in a sinusoidal type function.

The velocity of light is not the absolute maximum velocity, it is near the minimum, which is just above the energy amplitude for transformation from energy into mass. Stated logically "the velocity of light (*c*) is the average velocity for which energy and mass begin to change characteristics." Light velocity varies as a sine wave function.

The minimum velocity of light is the root mean square (RMS) of this sine wave variation. This velocity change continues at each specific frequency of light and the associated energy.

At light speed (c), or just below light speed, energy's velocity and frequency allows it to become mass (photon) part of its cycle; then energy the remaining part of its cycle.

A photon's maximum mass appears at zero velocity (inflection). Other normal actions that we are familiar with also oscillate in a sinusoidal variation, such as the electricity in our homes. This type action causes our lights to vary at 60 cycles per second (60Hz). Our eyes do not normally notice this electrical frequency variation.

When measurements of wave functions are taken (such as radio frequencies or the electric circuits in our homes) the instrument used is normally calibrated to measure the root mean square (RMS) of the sine wave. In other words, if your house electricity measures 110 volts, which is the RMS, the peak (PK) voltage would be 2sin 45° or 155 volts, and the average is about 98 volts.

The RMS of alternating voltage is used to calculate power. To maintain consistency the optimum function, which is the RMS of the energy sine wave, will be used for our calculations.

As the EME frequency increases, the wavelength time decreases, thus the time of mass would also decrease per wavelength (λ). The number of λ per second would increase, thus the amount of mass per second (not per wavelength) is dependent upon the associated EME frequency.

The velocity of visible light and other electromagnetic energies vary with this sine wave function, although some energy vibrations may have shapes other than smooth sine waves, such as a square wave or saw-tooth wave that are used in electronics.

Sine wave actions are continuously occurring throughout the universe. Strange particles (such as neutrinos) will appear and disappear depending on their velocities, and at the rate of visible light frequency and higher. Even our planetary orbits are somewhat *sinusoidal* rather than *elliptical*. This is explained later in Phase V.

For mass to remain mass it must loose its ability to reach c ($2.99792458E^{10}$ cm/sec). Also, if you look at it another way: ⬤⊂⬤ Wave propagation is a result of a force of generation (such as an antenna, heat, object entering water, or nuclear particles). If light is just a waveform then mass may be the generator. After all, energy, whatever it is, is also a mathematical concept describing the actions and characteristics of mass.

Velocity and temperature are not energy, they are indicators of energy. EME frequency is also an indicator of energy. The greater the energy the higher the frequency. However, this does not preclude a low frequency from possessing high "power." Heat is considered energy, but it is actually another indicator of energy exerted through molecular motion (vibration).

There is a distinct difference between mass, energy and velocity, although the only difference between mass and energy is velocity. The third variable in the formula "$E=mc^2$" is c. If a particle reaches the velocity of light, it becomes energy. Therefore, it is logical to state that if energy velocity dips below light velocity, it will become mass. This action may continue in a sinusoidal motion, as particles briefly appear at the energy sinusoidal wave inflections.

The actions occurring at inflections have no formal name, abbreviation, or symbol associated with them, so an action name must be excogitated (neat word, eh!). The **trans**formation from energy to mass, and the con**version** from mass to energy action will be called "**Transversion**," noted by "**Tv**," for convenience of this writing. This is to insure a clear distinction between transversion symbols and other scientific symbols such as "T" for temperature, "T" for tera, or "T" for Tesla.

If one believes Einstein in that mass cannot exceed c, then the mass part of an EME wave must be at a velocity below c. For that reason, the maximum amount of mass would be at the point of the sine wave inflection (zero velocity). This inflection is almost instantaneous, it is way too short a time for mass to appear and disappear, so the beginning of mass to appear should be somewhere below c and above zero. At c this would show that the peak velocity of a wave to be infinite. The answer to this is to buck Einstein's "nothing faster than light" theory and agree that the velocity of light energy can exceed c, but *only energy* can do this, not mass.

To plot the velocity of light energy/mass, let mass start at about 45° in this sine wave. This trace plot will appear similar to a cycloid with the cusps at zero velocity and the peaks (vertex) greater than c. Each wavelength contains two inflections and two peaks.

Calculations show that the total length of the energy part will be 0.7071 (sin 45°) of the wavelength; and the total length of mass will be 0.29289 (1-sin 45°) of the wavelength.

EME Mass and Velocities

In figure I.1 each EME wavelength (λ) generates two mass particles (photons), one at each inflection (centering around zero velocity). The total mass in one wavelength is ~ 7.3725E^{-48} gram (g) ($m=h/c^2$). The mass for one Tv is 3.68625E^{-48} g (1/2 the total mass in a wavelength).

It seems logical that the average velocity of the photon mass would be about 0.5 c (~1.498962E^{10} cm/sec), then the calculated average velocity of the energy cycle will be ~5.1178E^{10} cm/sec. The peak velocity of this energy wave is ~7.23763E^{10} cm/sec. This higher energy velocity compensates for the slower velocity of mass, allowing the average velocity of the total EME wavelength to be c (2.99792458E^{10} cm/sec).

The energy of one EME wavelength (1 Hz) is equal to Planck's constant (h), being 6.6260755E^{-27} erg sec. Each transversion (Tv) is one photon (3.68625E^{-48} g). There are two Tv's per wavelength (λ). Total mass per λ is 7.3725E^{-48} gram. See Figure I.1

Figure I.I

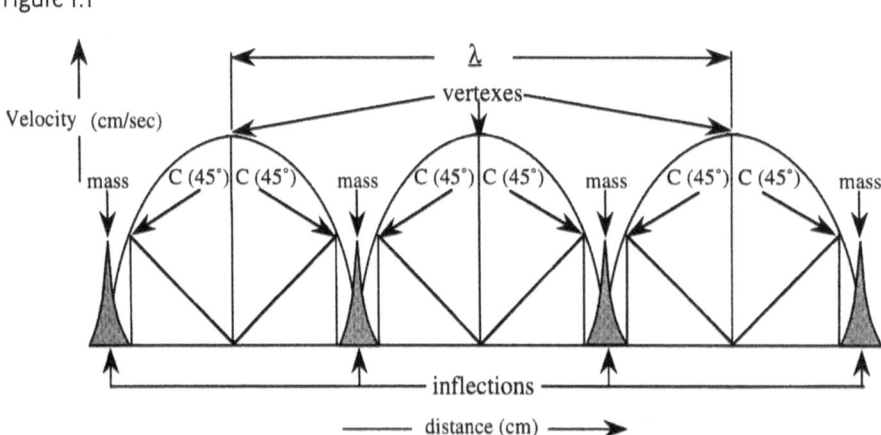

Illustration not drawn to scale.

The shaded areas show the mass transversion sections of wavelength. There are two transversions (Tv) in one wavelength. Each Tv begins at 45 degrees before inflection and ends 45 degrees after inflection. The velocity of c is at 45 degrees. Each particle at inflection is $3.68625E^{-48}$ gram, giving a total mass of $7.3725E^{-48}$ g/λ. The average velocity of mass is ~ 0.5c (~$1.5E^{10}$ cm/sec).

The mass that this energy produces per λ is calculated by:

$m = h/c^2 = 7.3725E^{-48}$ g.

Note:

h is $6.6260755E^{-27}$ erg sec.

c is $2.99792458E^{10}$ cm/sec.

g is gram.

λ is wavelength in centimeters (cm).

These calculations are derived from: E = **mc²**, and E = **hf**, therefore **mc² = hf**, or **m = hf/c²**. If the frequency is one Hertz (Hz), which is one λ per second, then **m = h/c² = 7.3725E⁻⁴⁸ g**.

The mass of **7.3725E⁻⁴⁸ g** is the total mass of one electromagnetic energy (EME) wavelength, regardless of frequency, energy, or velocity. Energy determines the frequency, frequency determines the number of λ per second, and λ per second determines the total amount of mass per second; but, not the amount of mass per λ.

Mass per second (m_s) is calculated by: $m_s = 7.3725E^{-48}$ g multiplied by frequency in Hz ($m_s = 7.3725E^{-48}$ g x f).

For example: If we have a frequency of $5.9E^{14}$ Hz (average visible light frequency), and we multiply $5.9E^{14}$ Hz by $7.3725E^{-48}$ g, we then get ~ $4.35E^{-33}$ g/sec, for that particular frequency.

Force of Light

If light has mass, then light will exert pressure. Or, if light exerts pressure then it has mass. Light pressure is measurable, it has been measured to be about $1.0E^{-5}$ dyne/cm² or ~ $1.5E^{-10}$ lb/in².

To calculate the amount of mass generated by the visible spectrum of EME, the average frequency between red light ($4.3E^{14}$ Hz) and violet light ($7.5E^{14}$ Hz) will be used. This average is ~ $5.9E^{14}$ Hz. If this average frequency is multiplied by $7.3725E^{-48}$ g/λ, the result will be ~ $4.35E^{-33}$ g/sec.

To integrate the EME range of visible frequencies from red to violate the average of $4.35E^{-33}$ g/sec must be multiplied by the number of frequencies included in this range. This is about $3.2E^{14}$ frequencies. Thus, $4.35E^{-33}$ g/sec x $3.2E^{14}$ frequencies = $1.3919E^{-18}$ g/sec for the entire visible light frequency spectrum of red through violet. The mass of $1.3919E^{-18}$ g/sec creates a measurable pressure of about $1.5E^{-10}$ lb/in².

If the EME that is bombarding us from space is included in pressure calculations it will show quite a force pushing on the earth and its inhabitants. There is much greater energy and higher frequency EME than $1.51E^{26}$ Hz (1 erg-sec), but 1 erg frequency ($1.51E^{26}$ Hz) will be used for calculating convenience. To calculate this force (pressure), the average frequency between 1.0 Hz and $1.51E^{26}$ Hz will be used, which is ~ $7.55E^{25}$ Hz. If this average is multiplied by $7.3725E^{-48}$ g/λ, the result will be $5.566E^{-22}$ g/sec. When this figure is multiplied by the integrated number of frequencies ($1.51E^{26}$ Hz) we will arrive at the total amount of g/sec received from space: $5.5662E^{-22}$ x $1.51E^{26}$, equals about $8.405E^{4}$ g/sec bombarding us from space.

We'll use a ratio formula to convert this to pressure. If light (red to violet) has a mass generation of $1.3919E^{-18}$ g/sec that exerts a pressure of $1.5E^{-10}$ lb/in², then $8.405E^{4}$ g/sec generated by the UOE will exert a pressure of about $9.0E^{12}$ lb/in². This is indeed, enough force to effect gravity on the earth, *et alii*. You would think that this pressure would smash us like a squashed grape, but this force is almost equalized by the energies coming from the opposite side of the earth, the energy that is not absorbed by the earth. The earth absorbs a comparatively small amount of EME to effect its "gravity." The explanation of the absorption of the UOE will continue in Phases III and V.

Gravity action takes place at a much, much greater velocity than light. Gravity propagates at least at $2.0E^{14}$ c and greater. Scientists have been searching for gravity waves. If a gravity wave (GW) is produced by an exploding star as scientists believe, the GW would arrive at earth long before visible light. One would be looking for the GW when it has already gone by.

We and the earth are absorbing about $3.0E^{33}$ erg/sec/g, coming from all directions. We are also emitting and giving off energy and mass from our bodies, but a little less than at the rate we absorb it. A small bit of the energy/mass we absorb (~ $1.0E^{12}$ erg/sec/g) remains with us as mass and energy, thus the UOE is truly a life giving "energy."

There is an effect called the "Photoelectric effect." This is electrons being knocked off of a surface by light. The intensity of light is not the direct cause of this effect. It appears that there is a threshold frequency that will begin knocking off electrons. The mass of an electron is ~ $9.1094E^{-28}$ g. It takes almost $2.5E^{20}$ photon masses to equal one electron mass. Therefore, it would take many, many photons to dislodge an electron. The higher the frequency the more photons per second there are chipping away at an electron. Also, the higher the frequency the greater the pressure (force) exerted on a body being bombarded by photons.

This is the reason the threshold for the photoelectric effect depends on frequency, rather than intensity. Light intensity does not affect the photoelectric action until the specific threshold frequency is obtained.

Please note that mass has intrinsic wavelength. The frequency (*f*) of this wavelength can be calculated by the UOE formula "$f = m_o/m_\lambda$," where m_o is the mass of the object, and m_λ is the mass of one EME wavelength. Briefly stated: the associated frequency of an object is determined by the number of photons creating it.

Following is a list of UOE symbols and formulae: Time (T) is in seconds (s). Length (L) is in centimeters (cm). Frequency (*f*) is in Hertz (Hz). Velocity (V) is in centimeters per second (cm/s). Mass (m) is in gram (g). Wavelength (λ) is in centimeter (cm). Transversion (Tv) is mass/energy sinusoidal characteristic exchange (two per λ). Electromagnetic energy (EME) includes visible and non-visible light.

Note: Formulae [I-1] through [I-7] are generally accepted standard formulae. This formula list presents the UOE symbols used in the **Dark** Side of Light:

f	standard frequency (Hz)	$= E/h = c/\lambda = mc^2/h$	[I-1]
c	light velocity	$= 2.99792458E^{10}$ cm/sec $= \sqrt{E/m}$	[I-2]
h	Planck's constant	$= 6.6260755E^{-27}$ erg sec (energy of 1 λ)	[I-3]
E	Standard energy (erg)	$= hf = mc^2 = hc/\lambda$	[I-4]
λ	standard wavelength	$= ch/E = c/f$	[I-5]
T_λ	time of wavelength (sec)	$= \lambda/c = 1/f$	[I-6]
V_λ	velocity of λ (avg.)	$= c = 2.99792458E^{10}$ cm/sec	[I-7]
L_m	length of mass per λ	$= \lambda \times (1 - \cos 45°)$	[I-8]
L_m'	length of mass per Tv	$= L_m/2$	[I-9]
V_m	velocity of mass (avg.)	$= \sim 1/2\ c = 1.49896229E^{10}$ cm/sec	[I-10]
T_m	time of mass per λ	$= L_m/V_m$	[I-11]
T_m'	time of mass per Tv	$= T_m/2$	[I-12]
m_λ	gram of mass per λ	$= h/c^2 = 7.3725E^{-48}$ g/λ	[I-13]
m_λ'	gram of mass per Tv	$= m_\lambda/2 = 3.68625E^{-48}$ g/Tv	[I-14]
m_s	gram of mass per second	$= f \times 7.3725E^{-48}$ g	[I-15]
L_E	length of energy per λ	$= \lambda \times \cos 45°$	[I-16]
T_E	time of energy per λ	$= T_\lambda - T_m$	[I-17]
V_E	velocity of energy (avg.)	$= L_E/T_E$	[I-18]
m_{ph}	photon mass	$= m_\lambda'$	[I-19]
dn_{ph}	photon density (g/cm³)	$= 2.2415E^{-45}/\lambda^3$	[I-20]

Universe of Energy Velocity Variations

The UOE's velocities throughout the universe, among other things, are mind boggling. We, as students, have been taught (and many have believed) that light velocity is "max ever fast." However, visible light is actually near the minimum velocity for electromagnetic wave energies.

Most EME in space, greater than 1 erg, has a much higher velocity than visible light, but we cannot measure what we cannot detect. Most energy maintains a high percentage of its state as energy due to its tremendous velocity and super-duper high frequency. A very small percent of this energy may go through many stages of slowing. As these energies are slowed, they "transform" to other types of energies until some are slowed to light speed. These energies vary between "mass" and energy at the rate of their wavelength near the frequency of light, as in figure I.1.

Please note that EME cannot be directly detected as energy. Energy is detected only during its Tv period as mass.

The only way we know energy is present is through the action of its mass. The following statement may be repeated in this writing. "Energy, whatever else it is, is also a mathematical concept describing the action and characteristics of mass."

Light to Dark

Energy may begin its transversion to mass at a slightly greater velocity than light or at a slightly less velocity than light. This "gray" area of transversion partially depends on the "medium" involved.

The gray area of mass/energy transversion is between c x sin 45° ($2.1198528E^{10}$ cm/sec) and c/sin 45° ($4.2397056E^{10}$ cm/sec). Within this gray area mass and energy may (or may not) remain as mass or energy depending on the energy amplitude, frequency, wave shape, and the surrounding medium.

Mass is either latent reactive energy (LRE) or as mass in transversion (m_λ'). LRE are EME particles that have interacted with other mass which then interferes with their return to full energy, velocity, or state. In this action they remain mass as LRE. Often, only part of a Tv mass may be acted upon during transversion, and only part of the photon may return to energy and part remains LRE. Both these actions are known as the "absorption of energy" by mass.

The longevity of a photon is dependent on the velocity and frequency of the associated energy. Of course, both, high or low energy photons' longevity may be altered through the interactions with other photons or LRE and/or the affecting medium photons.

EM energy with frequencies around visible light propagate at light velocity. Frequencies with energies much greater than visible light (about 250 billion times greater) begin to propagate at velocities much greater than light velocity. This reveals a slight discrepancy in energy/mass relation formula "$E = mc^2$." The point is that $E = mc^2$ may not always be accurate, and the velocity of EME is no where near the maximum speed limit our local physics police would have us believe (and ticket us for calculating otherwise).

Generally, the greater the energy the greater velocity, and the greater the frequency, therefore the less cycle time for mass transversion. During transversions the m_λ' that is moving from the decreasing velocity direction to the increasing velocity direction, and vice versa, creates an electrical "charge." This transversion action can also create other effects such as photovoltaic effect, photo-ionization, magnetic flux, and spin.

All matter is energy, but all energy is not matter. Mass does not absorb energy in the manner we may think, like tissue absorbing water. As energy Tv and mass (LRE) interact, part of the energy (m_λ') slows to less than c, thus becoming "mass," attached by the UOE to other mass. The "attached" mass remains mass, restrained from reaching c velocity. Mass is merely energy with characteristics we consider mass (density, weight, volume, etc.) It is actually the energy forces that create these mass characteristics.

Mathematically we consider energy only as a concept to describe mass. The only difference between energy and mass is velocity, subsequently when the EME velocity is at about c or greater it is energy. When EME velocity is less than c we see it as mass. As mass begins to transform to energy it loses mass characteristics and gains wave function characteristics as it approaches c.

The Compton wavelength (λc) = h/ρ, where λc is Compton wavelength, h is Planck's constant, and ρ (Greek "rho") is the particle's momentum (mass times velocity). These are some of the characteristics we measure and so determine whether we consider something as mass or energy. Remember, only the mass part of EME can be detected, not the energy part.

Note: The formulae "$E = mc^2$" and "$E = hf$" do not truly apply to normal everyday mass calculations. They are derived from characteristics of electromagnetic energies (EME). Actually, "$E = mc^2$" is derived from the algebraic combination of "$E = hf$" and "$\lambda c = h/\rho$." To calculate energy/mass/force for non-EME, the momentum and kinetic energy formulae are applicable.

History

A little history of physics should be reviewed, along with some background of a few important EME Physicists and their findings.

The French mathematician and astronomer Marquis de LaPlace, (1749-1827) had previously investigated and concluded that the energy velocity propagation of gravity is immensely greater than the velocity of light.

Sir Joseph John Thompson (1856-1940), British physicist (and father of Sir George Paget Thompson, another prominent British physicist), in 1881 established that when a charged spherical conductor is moving in a straight line it would perform as if it had extra mass of $4/3c^2$ times the electrostatic field. Later, around 1922, this was corrected by E. Fermi to $1/c^2$ times energy field.

Poincare, Jules Henri, a French mathematician and physicist, in 1900 suggested that electromagnetic energy may have density equal $1/c^2$ times the energy density ($E = mc^2$), E is energy, m is mass, and c is the velocity of light.

In 1905 Albert Einstein stated that if an object is losing energy (in the form of radiation), its mass will diminish by about $1/c^2$ times the amount of energy lost. He also, agreeing with Poincare, stated that the mass of a body is a measure of its energy content. Einstein stated that this is a sufficient condition to satisfy the law of conservation of motion. Later on, in 1908, G.N. Lewis utilized the radiation pressure theory to mathematically prove that an object absorbing radiant energy will increase in mass by the equation $dE = c^2 \, dm$, and declared that the mass of an object is a direct measure of its total energy according to $E = mc^2$.

Of course, an object will increase in mass when absorbing energy, but a photon will not increase in mass due to increase in velocity. A photon's mass will decrease as its velocity increases, and increase in mass as its velocity slows below c ($m = E/c^2$). Remember, energy can only be affected during the transversion period (at inflection) when it is mass. A photon is the mass part of EME. Photons are the building blocks of matter.

All objects in the universe are absorbing photons, thus are increasing in mass. These objects are also losing some mass through various actions.

The energies in the universe often conflict as mass during transversion periods. Sometimes this mass interaction emits special frequencies we know as light, and also some frequencies that we may or may not discern.

For many years there has been an ever steady increase in polishing and shining of old theories. This cosmetic appliance of physics has directed many brilliant people away from innovative concepts in science. Maybe it's about time for a face cleansing of our old "max faster" look, and look into a cleaned mirror image to be able to advance science at an un-powdered rate.

The "laws" of the universe are in harmony to the fine tune of less than one note in a googolplex. This means that the smallest finite value change affects a change in the largest infinite value.

We must remember even though energy and mass are one and the same (their individual characteristics are defined by velocity) they do not interact until their characteristics match as mass.

As energy increases so does the frequency of the associated electromagnetic propagation, however there is a point where frequency does not continue to increase linearly with energy. The basic cause of this is that the total mass per second becomes quite large causing the photons to interact, which increases the velocity of the propagation rather than the frequency. This very high number of wavelengths per centimeter make a crowded condition in the UOE which serves as an EME medium. When an increase in energy does not increase the frequency proportionally something must increase. That's where EME velocities begin to magnify. Further UOE energies will be addressed in Phase V "Astro Energy Levels."

The interactions of the transverted mass happens quite often at super high frequencies. The EME propagation velocities increase with the square root of the energy, but frequencies increase proportionally with the energy up to the "crowded" point. The energies of light support frequency increases up to about one erg ($1.51E^{26}$ Hz) then they begin supporting the velocity increase due to the "crowded" condition of photons.

I can tell you exactly how much mass is in the universe; if you tell me exactly what size the universe is. An answer that is somewhat near the correct amount may be given anyway (again, only a ballpark figure) for the "known" universe.

The total mass in grams, cycling on and off with the frequency, is greater than 3.55E raised to a googol (3.55 googolplex or $3.55E^{100}$ g) of our known universe (about 15 billion light years in all directions). The mass/energy density (MED) of the UOE is $> 3.0E^{19}$ gm/cm^3 for mass and $> 3.0E^{40}$ erg/sec/cm^2 for energy.

If we only consider a trillion light years from us in all directions as our known universe, this universal sphere will contain about $3.55E^{90}$ cm^3 with about $3.0E^{110}$ gm/sec of transverting mass. Just the major part of our solar system ($\sim 9.0E^{44}$ cm^3) alone contains about $3.0E^{64}$ gm/sec of transverting mass at any given microsecond. Although, there is infinite mass in our infinite universe.

Every cubic centimeter of the universe is jam packed with photons, very fast moving photons, whether they are EME as m_λ' or LRE as "permanent" mass. Some may even be considered neutrinos, although a neutrino has a much greater mass than a photon, yet many photons together can create a larger particle. There are very high energies (and energy velocities) in the universe, most are so high that we cannot detect them, only attempt to calculate them.

Light Wave Patterns (Interference Patterns)

To explain the association of the transversion period of a wavelength and interference patterns presented on a target, imagine a setup using a source of electromagnetic radiation (light), a barrier with an aperture to allow passage of EME, and a target such as a screen or reactive film. Similar experiments may have been previously performed, but none of these have been used to include and prove transversion mass.

Figures I.2 through I.8 present the patterns for various targets positioned at different wavelengths from the barrier aperture. Figures I.2 through I.6 are single aperture setups and figures I.7 and I.8 show double apertures setups. If target distances are lengthened there will be additional pattern rings due to the second, third, etc. wavelength transversions.

See Figures I.2 through I.8 (Illustrations are not drawn to scale)

Calculation of the radius of a fringe pattern can be accomplished using simple trigonometric functions. Figure I.4 presents a diagram showing the EME source, an aperture, a target, and one wavelength exiting the aperture at its Tv period.

Figure I.2

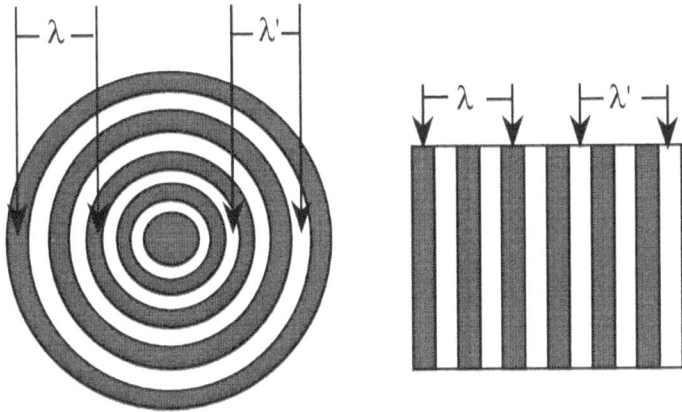

Interaction patterns are not caused by
interference, but rather by transversions.

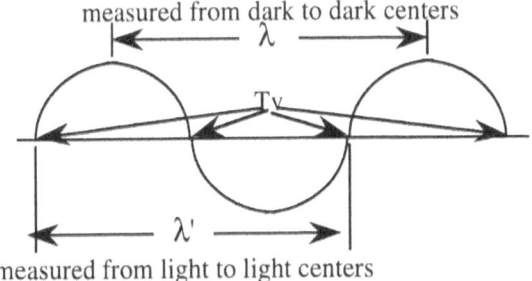

Interactions with matter will only occur when the electromagnetic
energy is matter itself during transversion.

Figure I.3

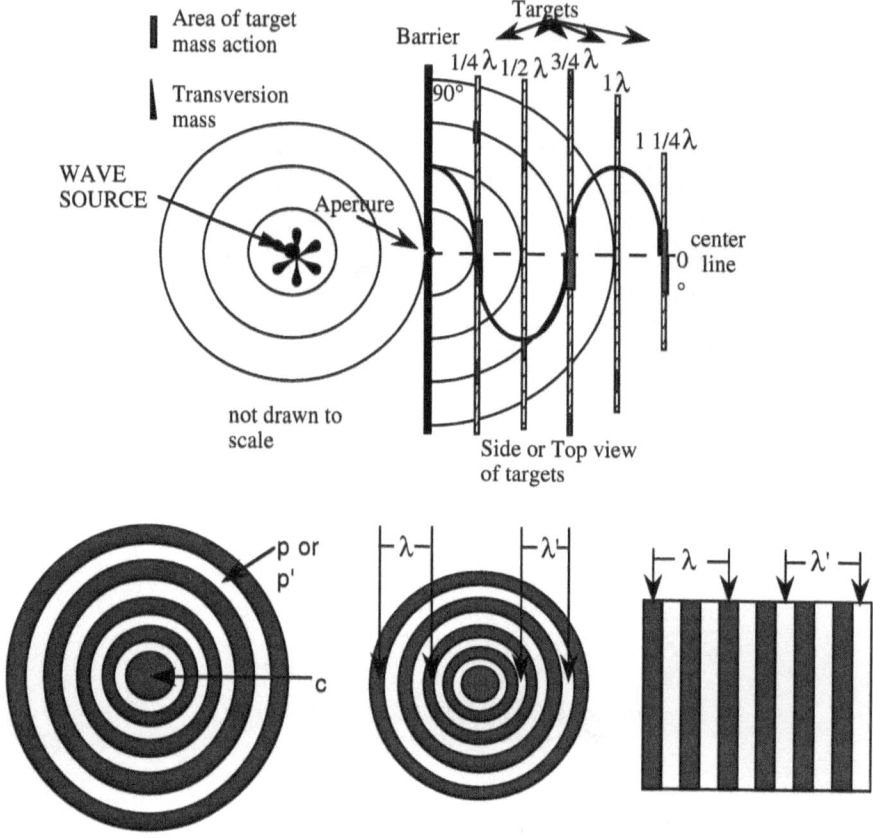

Figure I.3

Illustration not drawn to scale.

T - target (solid reflector, not mirrored).   B - Barrier (to omit all waves except through aper.).

a - aperture (opening in barrier).   c - Centerline (from aper. to center of target).

p - midpoint on pattern caused by 1/2 λ Tv.   p' - midpoint on pattern caused by 1 λ Tv.

A - distance from aper. to target along c.   H - distance from aper. to p or p'.

O - distance from c p or p' (O = H x sin Q) or (H x sin (arcsine(A/H)).

Q - angle from H to c (Q = arccos(A/H)).

<u>Note</u>: A and H are usually known as a distance of λ or partial λ.

Figure I.4

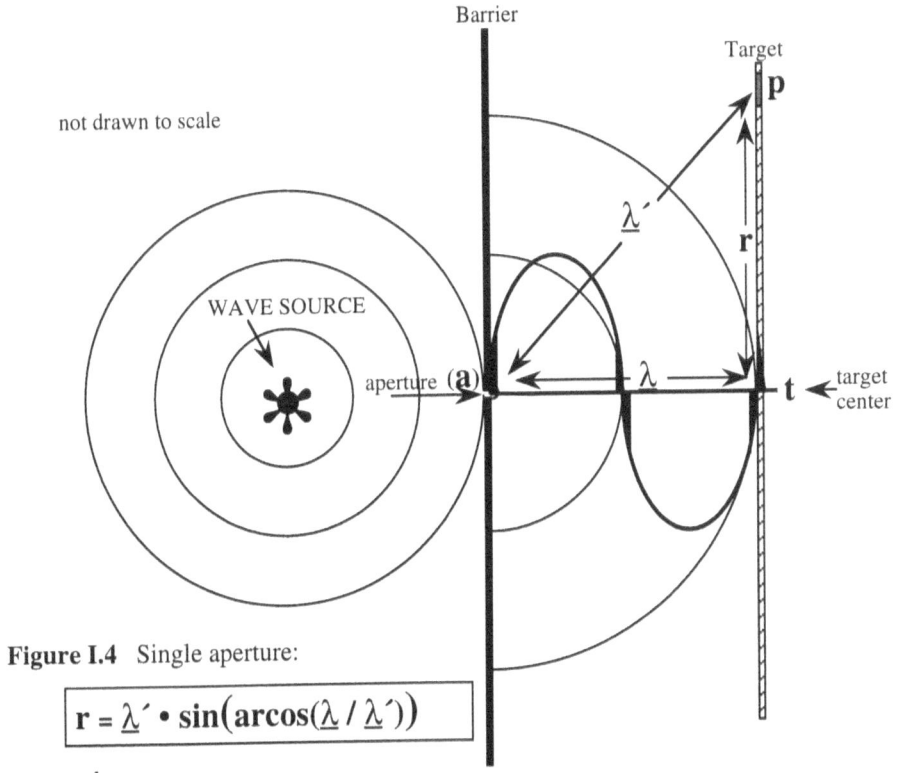

SINGLE APERTURE TARGET

Figure I.4 Single aperture:

$$r = \underline{\lambda}' \cdot \sin(\arccos(\underline{\lambda} / \underline{\lambda}'))$$

A is distance between a and t.
    a is aperture (round).
    p is pattern.
    r is radius of pattern.
    t is target.
    λ' is in multiples of .5λ, and must be greater than λ (e.g. 1.5λ, 2λ, etc)

Figure I.5

SINGLE SLIT TARGETS

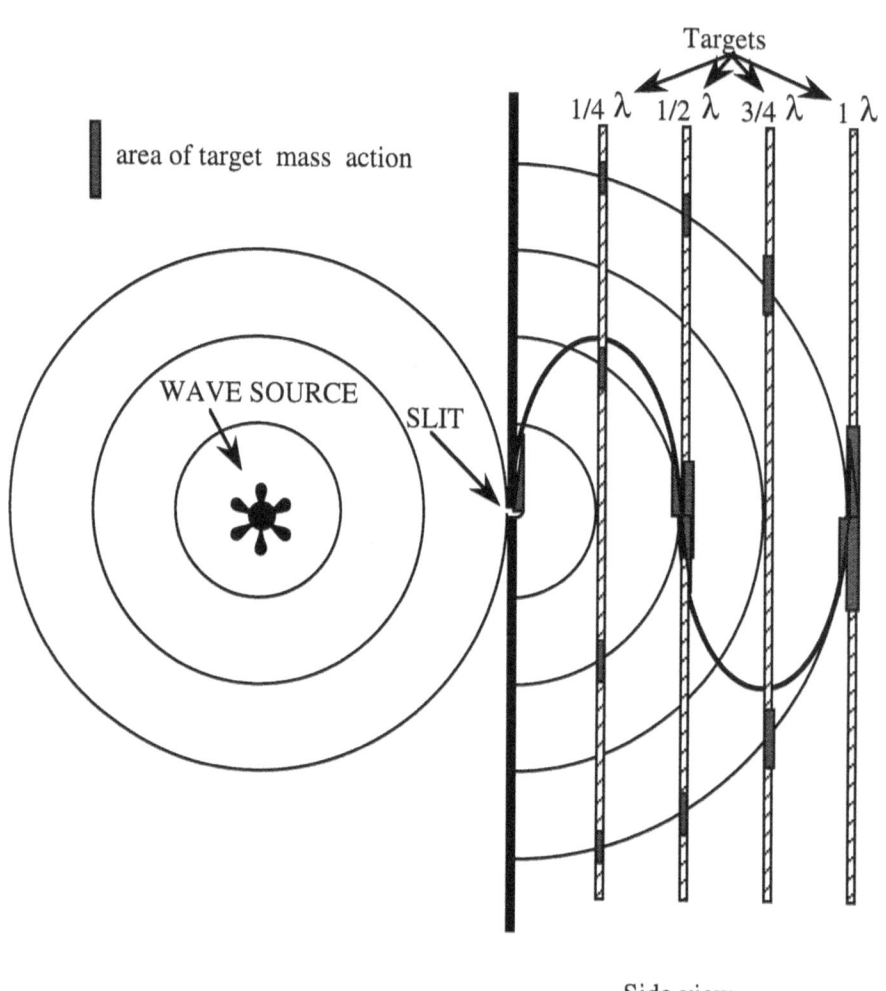

Side view
of targets

Figure I.6

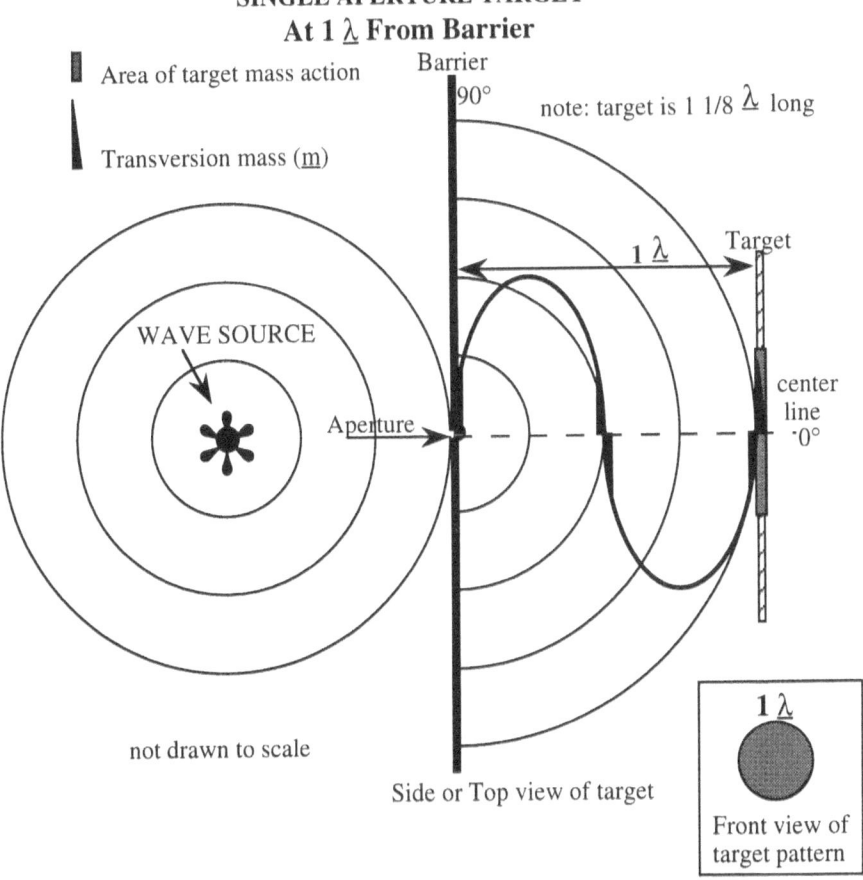

SINGLE APERTURE TARGET
At 1 λ From Barrier

Figure I.6
This single aperture barrier shows one wavelength starting at "zero" amplitude at aperture. The target is positioned at one wavelength distance from the barrier aperture. At this λ the wave propagation will be in transversion, thus the transversion mass will interact with the target as a single solid dot at zero degrees from the centerline and slit. The top of the barrier is at 90 degrees and centerline is zero degrees. Target is about one and one eight wavelengths long.

Figure I.7

DOUBLE SLIT TARGEtS

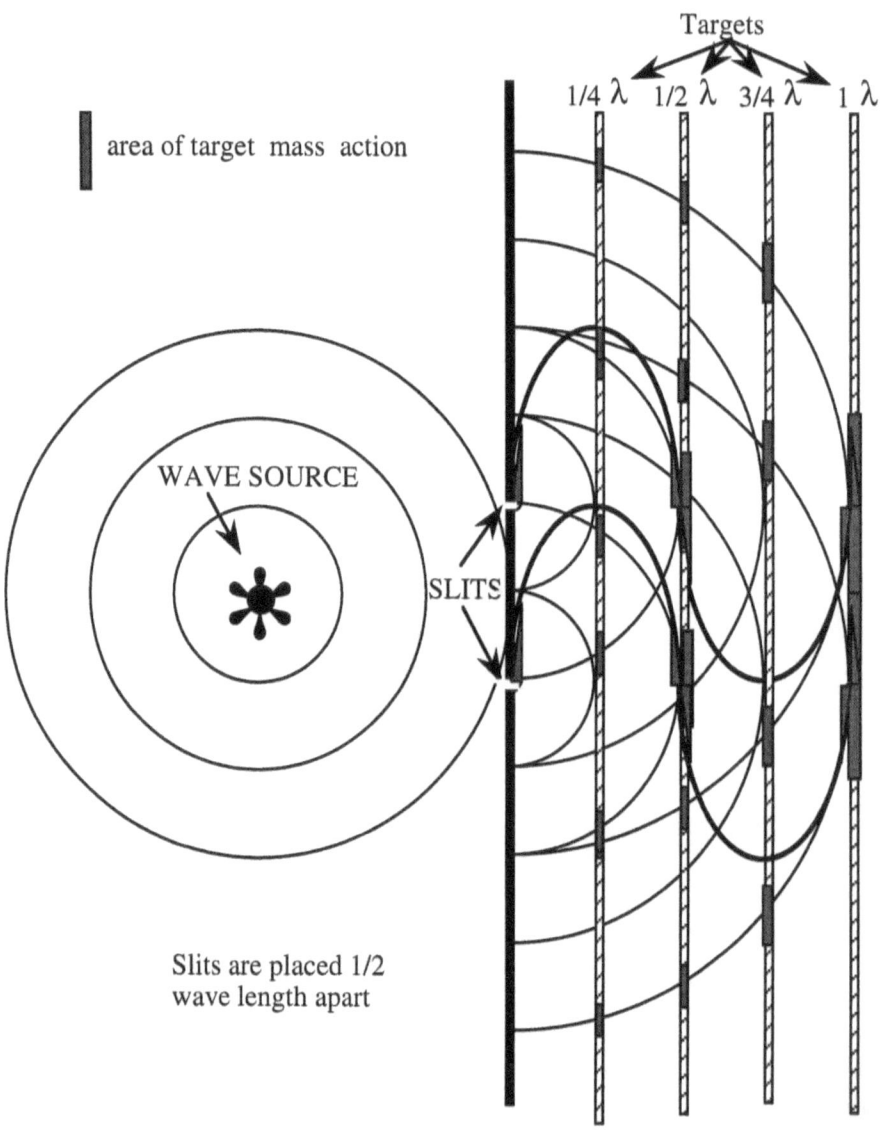

Side view
of targets

Figure I.8

DOUBLE SLIT TARGET PATTERN

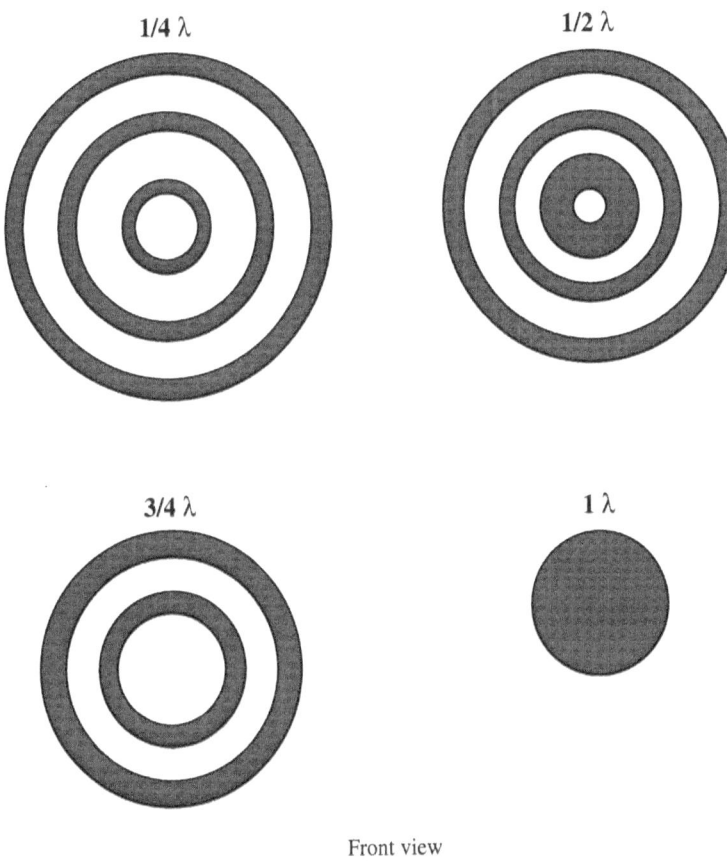

1/4 λ

1/2 λ

3/4 λ

1 λ

Front view
of targets

The pattern radius = λ' x sin (arcos(λ/λ')). Refer back to Figure I.4, page 41.

where: A is distance between aperture and target.

λ' is in multiples of .5λ, and must be greater than A.

r is distance from target center to transversion pattern ring radius.

If the EME wave exits the aperture at a null (not in a Tv period) λ' must be converted to

.25 λ + λ (or whatever factor is required to allow for the offset null).

Photons

The photon has an energy of about 3.313E^{-27} erg sec (.5h), and a mass of ~ 3.686E^{-48} g (.5h/c^2). The photon is the building block of mass. Particles, large and small, are constructed of photons. One gram of mass consists of 2.7128E^{47} photons. These photons were, at one time, associated with EME as transversion photons (m$_\lambda$') before being "absorbed" into mass as LRE. The EME is now contained in their mass (E = mc^2), as intrinsic energy.

The density of photons vary widely according to their associated wavelength. The length and volume vary with the wavelength, but the maximum mass of a single photon is always ~ 3.68625E^{-48} g, no matter what the energy, wavelength, or velocity.

The mass and volume of a photon grow at the same rate as its velocity decreases. The photon's radius at maximum mass is one half the mass length. The photon increases and decreases as a "sphere" during its entire transversion. The volume of the maximum size of this photon can be calculated with the simple sphere volume formula, [vol. = 4/3 π r^3].

To calculate a photon's radius (r$_{ph}$) the wavelength (λ) must be multiplied by 1-cos 45° (0.292893218813), then divided by 4, [r$_{ph}$ = (λ(1-cos 45°))/4]. For a ballpark figure, the wavelength can be divided by 4π to obtain the approximate photon radius. Once the radius is established we can insert it into the volume formula.

The following is an example:

wavelength = 5.0E^{-5} cm (visible green light).
photon radius = (5.0E^{-5} x 0.29289321881)/4 = 3.661165E^{-6} cm.
volume = 4/3 x 3.14159265 x (3.661165E^{-6})^3 = 2.05563742E^{-16} cm^3.
density = mass/volume = 3.68625E^{-48} g/2.05563742E^{-16} cm^3 = 1.7932E^{-32} g/cm^3.

A wavelength of 7.0E^{-5} cm (visible red light) would yield a photon density of only ~ 6.53E^{-33} g/cm^3. Comparing these two densities we find that density increases with higher energy and frequency (shorter wavelength), and decreases with lower energy and frequency (longer wavelength).

The reason that lower frequencies (such as sub-infrared light) appear to act as wave, more so than acting as a particle, is due to the lower photon densities. Subsequently, higher frequencies (violet and above) appear to act as particles more so than as waves. This, of course, is due to the higher densities, and higher rate of photons per second. The visible EME range of frequencies fall in the middle, acting sometimes as waves and sometimes as particles, depending on the study being performed. So, we find the higher the energy, and thus frequency, the shorter the wavelength, the higher the photon density.

The photon density of a 5.0E^{-5} cm λ is ~ 1.793E^{-32} g/cm^3. The photon density of a far ultraviolet wavelength (λ = 1.0E^{-8} cm) is ~ 2.242E^{-21} g/cm^3. A far UV photon is 125 billion times denser than an average visible light photon; yet, water is about 4.0E^{21} times denser than a far UV photon.

The point is that photon densities can vary greatly, depending on associated wavelengths, creating havoc among experimenters and empirical studies.

Let's review the characteristics of a photon (ph) or m_λ':

Photon energy (E$_{ph}$) ~ 3.313E^{-27} erg sec (1/2 Planck's constant—$h/2$).
Photon transversion maximum mass (m$_{ph}$ or m_λ') ~ 3.68625E^{-48} gram.
Photon density in cm^3 (dn$_{ph}$) ~ 2.242E$^{-45}/\lambda^3$.
Photon average velocity during Tv ~ 1.5E^{10} cm/sec (~ 1/2 c).
Photon average momentum ~ 5.5294E^{-38} g/cm/sec.
Photon mass per second (m$_s$) ~ m$_\lambda$ x frequency (Hz).
Particle frequency ~ particle mass (m$_p$)/m$_\lambda$'.
Particle wavelength (λ_p) ~ m$_\lambda$ x c/m$_p$ or λ_p = 2.21E^{-37}/m$_p$.
Notice that the total mass of a particle determines the frequency (or wavelength) of that particle. The greater the mass of an object, the higher the intrinsic frequency (and energy) of that object.

Review and Notes On The DARK Side of Light

There are two infinities, one going larger and the other going smaller.

If an object or particle has momentum, velocity, weight, spin, charge, etc. it has mass. There is no physical description that doesn't concern mass. Even energy is the mathematical description of the action of mass.

"Pure" energy does not affect other "pure" energy. Energy has no effect on other energies unless they are each in their transversion, or latent reactive energy (LRE) as mass. In simple words, energy does not affect energy or mass. Mass does not affect energy, mass only affects mass. This seems to be a wild statement, but if this logic is adhered to for the UOE, many wild phenomena will fall. Energies of all types must have a medium in order to propagate. Of course, the UOE is its own medium as energy in the form of EME; thus, there is always a medium present. The UOE and its infinite number of ephemeral particles (photons), is the "aether" which may, or may not, be the elastic aether theory.

The velocity of light does vary. It varies from 0.0 cm/sec to ~ 2.5E^{11} cm/sec in a sinusoidal fashion, and much faster at the higher energies (> 1 erg sec). This is why we have interference and fringe patterns, inertia, particles, dark matter, forces, photons, novas, orbits, gravity, weather, and life. These are all created through EME and its photons.

The UOE theories show velocities greater than light, but it does not pretend to accompany theories of time travel. Nor, does the UOE entertain the paranormal beliefs. Those concepts are left to others more experienced in left field and right field politics.

As we know, the UOE does not limit velocities to c. If time, space, distance, and existence are infinities, then velocity may also be unlimited. If we cannot comprehend infinity, surely we cannot, in truth, speculate its limits.

The real and practical minds must invent experiments for the wild ideas theorists dream up, thus Particle Physics will be left to intelligent people. It seems they have more names for particles than for insects. Of course, there are an infinite number of particles and characters we have not yet discovered.

The area of physics that is a real nightmare is plasma physics. As weird and complicated as it is, it will be in the near future, a major contributor to technology and advancement for science.

As mentioned before, $E = mc^2$ does not really apply to normal everyday mass. The appropriate formula for mass/energy not associated with EME is the momentum formula "mv." The difference between momentum and kinetic energy is .5c. Kinetic energy (KE) = $.5mc^2$). A simple explanation: Momentum energy is the energy it would take to stop a moving object. Kinetic energy is the energy it takes to move an object. When an object is stopped, the energy used to stop it is exerted only for the time it takes to stop the object. The energy used to move and keep an object in motion must be continuous. Therefore, normally, more energy is required to keep an object moving than to stop it, yet the energy used to stop an object must overcome the energy moving it.

The greater the mass of an object the greater the outside energy is needed to move this object. Actually, the outside energy and the object's intrinsic energy are at battle. Remember, the outside energy is also acting on this object's mass with EME mass (m_λ'). This is "energy versus energy, using mass." To do this, energy must be in a sinusoidal type function, such as EME.

As stated earlier, the formula "$E = mc^2$" is derived from sinusoidal energy (EME) formula "$\lambda c = h/\rho$" and "$E = hf$." Both these two formulae are associated with EME wavelength.

The photon is a very, very small piece of mass. In order to realize the size, a comparison to another particle shall be used for an analogy. If the Earth (with a mass of about $6.0E^{27}$ g) was a photon, then an electron would have $2.5E^{20}$ times the mass of the "earth photon." Relatively, this would be equal to about $1.5E^{48}$ g.

The ratio of this mass would be equivalent to 750 trillion (750,000,000,000,000) suns. That's a lotta sunburn.

If our sun is an average star with the mass of about $2.0E^{33}$ g, and there are about 200 billion stars in an average galaxy, then the mass of $1.5E^{48}$ g would equal about 3,700 galaxies.

Of course there is no true "average" star, or average galaxy, but you get the picture, showing the size comparison of a photon to an electron. And we thought the electron was small.

A single photon cannot be detected. One wavelength per second (1 Hz) has a length of 186,000 miles. The length of m_λ' (photon) would be around 27,239 miles. This photon, indeed would have a very small density. The density of this photon would be about $4.355E^{-62}$ g per cubic mile. There is no way, right now, we can detect this single photon.

The super high frequency of $2.291E^{25}$ Hz would generate a photon density of only about one gram per cubic cm, and this photon would last just $6.4E^{-27}$ seconds. This frequency would also generate $1.69E^{-22}$ g/sec.

Infinity, Velocity, and the Future

As stated earlier, only energy can go faster than c, not mass. This is true up to about one erg frequency ($1.51E^{26}$ Hz). Around this level the energies begin supporting energy/mass velocity as well as frequency.

The UOE velocities, even though they are much greater than the speed of light, have been presented, mostly with limitations. There are two primary reasons for this. One is that most people, especially scientists, tend to stay away from infinities, for good reason. Except for integrals, etc. infinities tend to corrupt calculations, and usually state that something is wrong, or at least not all correct. The other reason is that the fabric of scientists, and all others, are not designed to comprehend infinity. This does not say infinity does not exist, it certainly does, and this we understand.

Nature has posted "No maximum mph—Drive Safely" signs, but we do not yet read them, for some scientists have posted c as a speed limit, and many others just string along, reluctant to buck science city hall.

If you could move faster than the speed of sound and then stop you could listen to the noise you've created in the recent past. You cannot change the past, at best you can only witness it. This logic also holds the same conditions for light. If you could travel faster than the speed of light and then stop, you could see the light arriving from the past, including the reflected light from the start of your journey, thus you'd see yourself arriving. Although, you would only see the reflected light and not your true self. You could affect the light, but not the past.

If you moved along with this reflected light you could alter its construction a wee bit. Then, if you took off again faster than light and once again stopped to look back you would see the altered light when it reached you. Only the light was altered, not the past. The visual appearance of the past only would be changed. This light perturbation would give a future witness a false impression of the past. We see this often from the stars, from the light being altered through the vast distances in space.

Although, at the present we cannot move ahead of light and look back, we can see light in a reflection which is light from the past, the very recent past. The time that light takes to go from you to the mirror and back to your eyes is almost insignificant, but it is a time frame. This means that we are a little older than what we see in the mirror.

Imagine yourself in your Indestructible Infinitely Variable Velocity SpaceShip (IIVVSS or (IVS)^2), on the way to a far star about 20 billion light years away. At the start of this journey you would see this galaxy's light which was emitted 20 billion years ago. As you move nearer to the galaxy this time span decreases. As you get half way the time span has been reduced to 10 billion years. Finally, as you reach this galaxy in your (IVS)^2, the time span for its light to reach you is close to zero.

With concern for the future and past, it matters not how fast you got to this galaxy, it appeared that you were closing the time gap from 20 billion years ago to the present, but this is not really so. What you were closing in on, of course, was distance.

A mathematician with design could twitch his magic pen and perform some "Costello" math to "prove" you went into the future, or past. Like the 5th, 6th, 7th, etc. dimensions, yet it is only math. Mathematicians can delve in many dimensions, but for practical and technical purposes we shall use only three, plus time. We must remember statistics are considered math, and we all know what a crafty liar can do with statistics. If you don't know, ask a few politicians, they'll be able to explain it in a roundabout way—like the cha-cha, a two step shuffle, or a waltz around.

Moving at any velocity from 1 mph to infinite mph changes not time, but the time, speed, distance ratio. Velocity does not change time, but this does not say that time (future or past) cannot be traveled, only that velocity is not the major mechanism to do it.

There are claims of occasions that people have seen or witnessed the future. If these claims are true then there are mechanisms other than velocity to span time, perhaps dimensions.

Time is a constant, life depends on it. Flawed empirical studies embrace a false theory of time and space warp. It seems that the more confusing a science theory is to the layman the more scientists embrace it. It is a little egotistical on many self-centered scientists.

A major part of the UOE theories are based on Planck's constant of 6.6260755E^{-27} erg sec. If this value is changed, then the UOE values must be corrected. Planck's constant is also a major value in deriving "$E = mc^2$."

As we know "$E = mc^2$" is derived from two other formulae: $\lambda c = h/\rho$ and $E = hf$. Where λc is the Compton wavelength, h is Planck's constant, ρ is momentum (mv), and f is frequency. Both these formulae are directly associated with electromagnetic energy. Therefore, $E = mc^2$ is really associated with EME, and not so much with everyday velocity of mass. Again, the kinetic energy and the momentum formulae are more acceptable for non-EME mass calculations.

Phase II
The Finite Universe

To begin our trip into another small world, we will treat our atom mathematically, but not with too much math. In order to start our calculations we must first discuss the information concerning the atomic particles we will be playing with. Mostly, these particles will be the electron (e-), the proton (p), the neutron (n), the atom nucleus (nuc), and the averages of the neutron and proton combined (np).

The average np is not a real particle, but we will use it to shorten a few of our calculations. Instead of dividing a nucleus into protons and neutrons for separate calculations, the average will be used. This average is normally a slight bit low, but it makes calculations much faster, easier, and results are still in the ballpark, between third base and home plate (although, you may think they're out in left field).

As you probably know, mass, energy, spin, etc. values for particles will vary from institution to institution, and experiment to experiment; therefore, the generally accepted values among them will be used. It appears that the atomic size particles may not absorb energy at the same rate as larger materials. No one knows exactly what rate electrons, protons, neutrons, etc. absorb the energy of the UOE, but an educated wild guess is about $1.0E^{12}$ erg/sec/g (about the same as human mass). For now this level fits in nicely and is used easily in calculations for orbital velocities. Eventually, as experiments and empirical studies evolve, corrections will be introduced.

If the total absorption of energy by an atomic nucleus was greater than $1.0E^5$ erg/sec, in many instances, it would cause the inner electrons to exceed light velocity. For now, the absorption of $1.0E^{12}$ erg/sec/g is sufficient for calculations.

Some general information has been pre-calculated for convenience. Listed below in Tables II-1 and II-1A are a few general descriptions of particles.

See Tables II-1 and II-1A

Table II-1

	electron (e-)	proton (p)	neutron (n)
mass	$9.1094E^{-28}$ g (me-)	$1.67252E^{-24}$ g (mp)	$1.6748E^{-24}$ g (mn)
density	$9.72E^9$ g/cm^3 (de-)	$9.72E^9$ g/cm^3 (dp)	$9.72E^9$ g/cm^3 (dn)
radius	$2.8177E^{-13}$ cm (re-)	$3.4504E^{-12}$ cm (rp)	$3.452E^{-12}$ cm (rn)
volume	$9.371E^{-38}$cm^3 (vole-)	$1.72E^{-34}$ cm^3 (volp)	$1.7E^{-34}$ cm^3 (voln)
surface	$9.977E^{-25}$cm^2 (sure-)	$1.497E^{-22}$ cm^2 (surp)	$1.5E^{-22}$ cm^2 (surn)
Eabs ~	$9.1094E^{-16}$erg/sec	$1.67E^{-12}$ erg/sec	$1.675E^{-12}$ erg/sec

Table II-1A

	neutron,proton (np)	1H¹ nucleus (H¹nuc)	92U²³⁸nuc (238nuc)
mass	$1.67366E^{-24}$ g (mnp)	$1.67252E^{-24}$ g	$3.984E^{-22}$ g
density	$9.72E^{9}$ g/cm³ (dnp)	$9.72E^{9}$ g/cm³	$9.72E^{9}$ g/cm³
radius	$3.452E^{-12}$ cm (rnp)	$3.45E^{-12}$ cm	$2.139E^{-11}$cm
volume	$1.72E^{-34}$ cm³ (volnp)	$1.72E^{-34}$ cm³	$4.099E^{-32}$ cm³
surface	$1.497E^{-22}$ cm² (surnp)	$1.496E^{-22}$ cm²	$8.479E^{-18}$ cm²
Eabs ~	$1.67E^{-12}$ erg/sec	$1.67E^{-12}$ erg/sec	$3.98E^{-10}$ erg/sec

Orbital Electrons

The orbital velocity of an electron depends upon the nucleus mass, the distance from the nucleus, other "nearby" nuclei, protons, electrons, etc., and the strength of the affecting UOE.

To play with a few orbital electron energies, the uranium 238 atom will be used. This atom has many orbital electrons, but for calculation simplicity the inner electron (ine-) and an outer electron (oute-) orbital energies and velocities are only used.

Refer to Table II-1A to find the energy shielded (Eabs) by a U²³⁸ nucleus. This is $3.984E^{-10}$ erg/sec ($1.0E^{12}$ erg/sec x $3.984E^{-22}$ g). The distance at which this shielded energy will be calculated is by the radius of the nucleus ($2.14E^{-11}$ cm). The distance from a U²³⁸ nucleus to the inner orbiting electron (ine-) is calculated by $1.0E^{-6}$ times the cube root of the nuclear radius (rnuc), is $1.0E^{-6}$ x (rnuc)1/3.　　　[II-1]

The distance from this same nucleus to its outer orbiting electron is calculated by $1.0E^{-4}$ times the cube root of the nuclear radius ($1.0E^{-4}$ x (rnuc)1/3.　　　[II-1A]
Note: These two formulae (II-1 and II-1A) are rules of thumb derived from the average of low atomic number to high atomic number isotopes.

To calculate the energy at ine-, the old inverse square law trick is used:
"$E_2 = E_1$ x D_1^2/D_2^2" is our basic inverse square law.
Where:

　　E_1 ... is $3.984E^{-10}$ erg sec
　　D_1 ... is nucleus diameter ($4.278E^{-11}$ cm)
　　E_2 ... is distance from the nucleus to orbiting electron (ine- = $2.776E^{-10}$cm)

　　$E_2 = 9.461E^{-12}$ erg sec for ine-.　　　[II-2]

To calculate oute- orbital energy, the D_2 distance is replace by $2.776E^{-8}$ cm, and the result is:

　　$E_2 = 9.461E^{-16}$ erg sec for oute-.　　　[II-2A]

If the kinetic energy formula (E=1/2 mass x V2) is transposed, one may discover orbiting electron velocities. For ine- we find an orbital velocity of:

$$\sqrt{(E/.5 \times m_e\text{-})} = \sqrt{(9.461E^{-12}/.5 \times 9.1094E^{-28}g)} = 1.44E^8 \text{ cm/sec (3,200,000 mph)}.$$

This is about 1/200th the velocity of light.   [II-3]

This is a very fast pace for a little ol' electron, but it's built to hack it.
The orbital velocity of oute- is: $1.44E^6$ cm/sec (32,000 mph).   [II-3A]
The outer electron orbital velocity is much slower than the inner electron, but still a very fast pacer.

For general information the minimum and maximum orbiting electron energies and velocities should be discussed. To do this the maximum energy shielded by a large, non-manmade nucleus (U^{238}) is used. The maximum velocity and energy have already been calculated above in formulae [II-2] and [II-3].

All there is left to calculate is the minimum electron energy and velocity. To do this, we need a nucleus that shields the least amount of energy to allow electron orbiting. Of course, this would be the hydrogen atom nucleus ($_1H^1$), consisting of only one proton for the nucleus and one orbiting electron.

The distance (D_1) to use is 2 x nucleus radius ($3.45E^{-12}$cm) = 6.9^{-12}cm. And the second distance (D_2) is the distance to the outer electron which is:
$1.0E-4 \times (3.45E^{-12})1/3 = 1.511E^{-8}$cm. The proton mass energy absorption is $1.6748E^{-12}$ erg sec (from Table II-1). Substituting these values into the inverse square law formula for minimum orbital energy: $E_2 = 8.72E^{-20}$ erg sec.   [II-5]
The velocity of this minimum energy electron is $1.384E^4$ cm/sec.   [II-5A]

Below is a list of the minimum, maximum electron orbital energies and velocities.

maximum e- orbital energy ($9.461E^{-12}$ erg sec)	([II-2])
maximum e- orbital velocity ($1.44E^8$ cm/sec)	([II-3])
minimum e- orbital energy ($8.72E^{-20}$ erg sec)	([II-5])
minimum e- orbital velocity ($1.384E^4$ cm/sec)	([II5A])

There are many orbiting electrons between the minimum and maximum energies, along with their various velocities, in all atoms and isotopes.

Comparing the minimum and maximum electron velocities by converting velocities to miles per hour: ~ 310 mph minimum to ~ 3,200,000 mph maximum, gives a clearer picture. Using these units, the wide range of electron orbital velocities are well recognized.

Atom Shells

Some physicists believe that atoms appear as clouded shells. This is probably correct. Clouded shell appearance can be easily explained. If the radius of an atom is ~ 5.3052E⁻⁸ cm (including orbital electrons) the distance that one of the orbital electrons will travel in an orbit is about 3.333E⁻⁷ cm.   [II-6]

If this electron travels at the rate of 1.0E⁶ cm/sec, it would be circling the nucleus at 3.0E¹² orbits every second. This is 3,000,000,000,000 revolutions per second (rps), or three trillion rps.

But why do atoms have "shells"? To answer this for our fair friends, another analogy will be used: ⬤◯⬤

If a ball on a string was being twirled around slowly (about 10 mph) and observed for a moment, the observer could see the variations in the ball's orbit, the dips and rises, and where they occur. If this same ball was twirled 5 times faster, the observer would see only larger integrated variations in the orbits (about $1/v^2$ variations). If this ball was at the end of a string 5 feet long, the distance the ball would travel in one orbit would be about 31.42 feet per orbit.

If the ball were twirled at 310 mph, it would be completing 868 revolutions per minute.

This seems too fast for the eye to keep up with, to observe the variations in orbit, but large variations will show as a moving blur. At 10,000 mph (28,000 rpm) the orbital variations would integrate into a blurred line, appearing as a solid ring. If this ball would be moved one degree sideways each orbit, it would also be completing about 78 rpm sideways. And if this ball were to rotate on its axis at about 300 rotations per orbit, it would be rotating at 8,403,300 turns per minute.

If this ball were twirled at 5.0E⁵ cm/sec (way less than an electron's average velocity around a nucleus) it would appear as a solid cloud surrounding the nucleus. Any observing instrument would be unable to discern the frequency of this moving electron.

At an electron's slowest average orbital velocity of ~ 1.384E⁴ cm/sec (~ 310 mph), just one orbit will take only ~ 6.86E⁻¹² seconds; or this electron will complete 1.46E¹¹ (146 billion) orbits in one second.

Even at "slow" orbiting electrons appear as clouds. That is why orbiting electrons' positions and states are "impossible" to determine at the same time.

See Figure II.1

Figure II.1

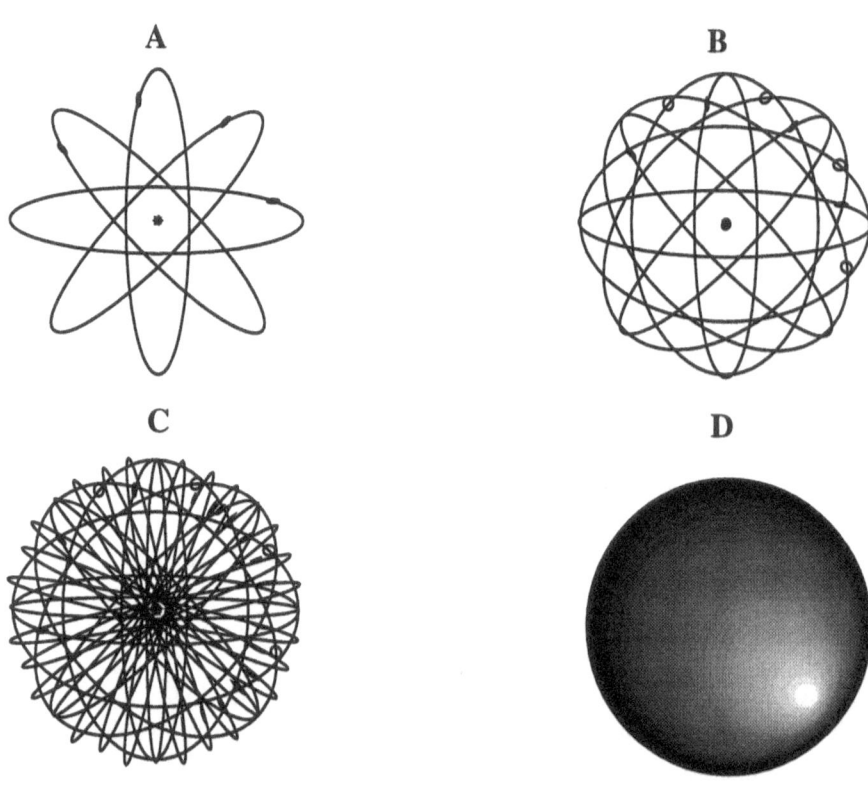

In Figure II.1, figures A, B, and C could only exist if the orbiting electron could be slowed down to just a few miles per hour. Figure D is the depiction of a normal orbiting electron.

An electron orbiting at 310 mph orbital velocity (minimum for an orbiting electron) would show its "path" as solid lines (figure D). As shown, these lines would blend into each other causing the appearance of a shell as the electron changes orbital directions each orbit. This change of direction is very, very small for each orbit, but even at the "slow" orbital velocity of 310 mph the orbits completed will be "astronomical." About $1.46E^{11}$ (146,500,000,000) orbits will be completed in one second.

When external energy is applied, such as for observation, the electron will react as though it has enough energy to exceed light velocity (which is not impossible).

If an instrument utilizes about $9.00E^{-4}$ MeV in an attempt to analyze an electron with an amplitude of 1.0eV, it applies about 900 times the subjects amplitude. The resultant observation is not an observation of nature, but of man's interference with nature. The electron and its associated wave packet will disappear into "pure" energy. This phenomenon can occur in an orbiting electron, or a "free" electron.

A type of strobe observation may "sync in" the electron every one billion orbits or so, but any applied energy from this strobe would have to be less than one percent of the electron's energy, so as not to affect its state.

The paradox we have to analyze an electron is that the lower the energy the longer the wavelength. The energy should be low, but the wavelength should be short with respect to the item being observed. So we have a slight experimental problem to solve, having the shorter the wavelength the higher the energy.

Any dynamic type measuring equipment's energy should be, for optimum observation, at most, one percent of the energy of the item being measured. A volt meter, for instance, should not redirect or absorb more than one percent of the voltage being measured. Our measuring devices are not quite sophisticated enough to accurately measure photons, quanta, electrons, etc. without causing an effect on them of at least one hundred percent. An instrument designed to be effectively efficient to measure quanta must also measure itself. Measurements of very high energy particles are quite feasible, and are performed with fair accuracy, but there are also energies that are too large to measure, or even detect with our present technologies.

If one would attempt to measure a photon using an electron, the results would be disastrous. Although, to measure an electron using a photon would be almost feasible. There are ~ $1.23E^{20}$ photons in an electron.

When a measuring device applies about the same amount of energy as the particle or photon being measured, the results are two hundred percent erroneous. This action would be like slamming Venus into the Earth just to measure Earth's velocity (or what it used to be). The other extreme would be trying to bounce a BB off Jupiter to measure Jupiter's reaction. This would yield too little energy to detect or even accurately calculate, plus the BB would be disintegrated by the giant gaseous planet.

Observers often interfere with the subject even before the observation. If a collapse of a wave packet has been observed, it is assumed the collapse is probably from the observation. The observation should be observed, although there is a quantum chance of fifty/fifty that the wave packet being observed would collapse even if it was not interfered with by the observation.

A single photon wave packet travels in all directions. A signal sent to a wave packet simultaneously reaches the different parts of this packet. This phenomenon causes the appearance of something being in two places at the same time (electron super state). This phenomena is occurring continuously.

The electron has an intrinsic frequency (wavelength) which is determined by the number of photons creating it. This can be calculated by the UOE formula $f = m_e/m_{ph}$, where me is the electron's mass and m_{ph} (or m_λ') is the mass of a photon (one photon is $3.68625E^{-48}$ g, two photons in one wavelength are $7.3725E^{-48}$ g total).

An example is:

$$f = m_e/2m_{ph} = 9.1094E^{-28} \text{ g}/7.3725E^{-48} \text{ g} = 1.23559E^{20} \text{ Hz.}$$

Note that the frequency is 1/2 the number of photons (two photons per wavelength).

The randomness accounted for in atoms and nuclear decay is a direct result of the UOE. When observing randomness, the observer may contribute to the chaotic results by physically contributing to the shielding of the UOE by being near the object observed. Mass "variation," the time of day, night, or even year, may have important effects upon many delicate, finite measurements. The direction or orientation with respect to east, west, north, south, and even up or down, and elevation may also affect experimental results.

Remember, determinant is through intelligence, randomness through ignorance.

An electron orbiting a nucleus at $2.9E^6$ cm/sec for one second, is compared to the earth orbiting the sun for more than a trillion years. The orbits compare, but not exactly due to the amount of energies affecting each one. These energies vary with different frequencies and wavelengths with respect to size.

Time Variable

If you would like to stretch your imagination and picture an electron as a planet similar to earth, we would see this electron being created, evolving to support life and returning to energy within earth time of around 1/2 hour. Electrons can certainly exist longer than 1/2 hour, and sometimes less. Many factors (almost an infinite amount) determine a particle's lifetime as mass.

Time is consistent, the "concept" of time is not. Time appears to be a constant, the value of it is a variable. The concept of time is dependent upon the "size state" of the observer.

To continue our imagination stretch, let's say we have two inhabited planets, one called Earth and the other called Lecton, with their size difference as comparable to earth and an electron, respectfully. Earth mass size would be 1.0 and Lecton's size would be about $1.5E^{-55}$. Earth's orbital velocity is about $3.8E^6$ cm/sec. Lecton's orbital velocity is only $5.0E^5$ cm/sec. Earth must travel $9.4E^{13}$ cm (total orbital distance) in $3.2E^7$ seconds (1 Earth year) to complete an orbit.

Lecton must travel $3.325E^{-8}$ cm (total orbital distance) in $6.65E^{-14}$ seconds (one Lecton year). The Earth travels about $2.98E^6$ cm in one second. This is only 0.000003125% of Earth's total orbit, but in this same second Lecton has completed $1.5E^{13}$ orbits. In reference to someone on planet Lecton, their time elapsed would be 15,000,000,000,000 (15 trillion) years to one of Earth's seconds (considering each orbit is a planet's year). Once again, a planet the size of Lecton may be created, generate life, history, and be annihilated, all within 1/2 hour of Earth time. If this is the case, time is relative to size, or at least "lifetime" is.

This is also true in earth's nature. Small living forms generally have shorter life spans than larger ones. Of course, nature usually provides some exceptions for us, although on a bell curve this lifespan to size is revealed.

The values of time are as varied as seconds in eternity. The value of time is dependent on the situation. For example: how long a minute is, depends on which side of the bathroom door you're on.

There are no two atomic particles exactly alike, including electrons. Atomic particle dimensions, along with time values, are as varied as the planets and stars.

Globs

The atom's nucleus is not composed of distinct separate particles. A nucleus is a ball of plasma with internal reactions. A silly likeness of these reactions, when split, would be as compared to a glob of jelly. If this jelly glob was slapped with a spoon, it would split into smaller jelly globs. None of these globs would be identical to each other. Also, if these globs were in space they would eventually become spheres, unless they were absorbed by other particles of different densities. Liquids in zero or micro gravity tend to become a sphere due to the UOE.

All particles such as neutrons, protons, etc. eventually "decay" due to the back and forth transformation from energy to mass and vice versa.

If particles can become energy, then energy can become particles ($E=mc^2$). There is no true "half life" of a particle. An infinite number of factors (including observations) determine the transformation from mass to energy (and vice versa) of all size particles. The naive abstraction of a specific non variable half life for particles will lend work to anyone searching.

There is no specific range for a particle to transform to energy. This is due to the amplitude resonance for each and every particle within itself. Particles in our own bodies transform back and forth due to an infinite number of factors. Our lifetime is controlled by energies and particles that are continuously intertwining and changing.

We have the technology to slightly alter this action to our favor, but nature will always win. Nature out figures us because it does not rely on mathematics. Nature calculates by nature, not math.

Analysis and Analogy of a Particle's Spin

A softball will be an excellent depiction of a nuclear particle that spins. Mark a small dot on this "particle." Place this mark near the center of the "particle" facing you.

There are two ways to rotate this "particle" 360 degrees horizontally, and simultaneously, 180 degrees vertically. The dot should be opposite you. If you repeat these moves, the dot will return to its original position. Now rotate this "particle" again, starting with the dot facing you, but we'll be a little more skillful this time, having practiced. Rotate the "particle" 180 degrees horizontally, and 180 degrees vertically, simultaneously. If you dropped it, then it could be called the Heisenberg Uncertainty Principle. When the particle is finally spun without inducing the Heisenberg Uncertainty Principle, the dot should be again facing you.

The actions of a rotating particle in two different directions simultaneously can produce many various combinations of spin actions and results. This analogy demonstrated the action of an atomic particle that must make two complete spins to return to its original position.

This analogy can be transformed (or extrapolated) to use for other spins such as 1/2, 3/4, etc. for a particle to return to its original position. Mathematical literal descriptions are not always in concert with the beauty of math.

Atom Nucleus

Most folks and scientists seem to picture the atomic nucleus as a compound of ball shaped protons and neutrons held together by the "strong" force. The nucleus is actually a conglomeration of photons joined together in a sphere, resembling the sun. This nucleus continuously absorbs photons, being heated as are the sun and earth, and all particles that "absorb" mass (energy). The actions of the solar system and an atom are very similar.

When an object is heated, its molecules vibrate. The more heat, the greater the vibration of the molecules. This molecular motion is created from the photon's actions. In an atom, of course, there are no molecules, yet an atom can vibrate. This motion is also created from photons. Although, most of the sun's atoms are plasma atoms and have no orbital electrons, they still consist of photons as do other atoms.

Heated particles may be known for their molecular motion, but the characteristic of the atom's nucleus should be known as photonic action. The sun's actions and characteristics are truly photonic actions.

All energy of any type is derived from photonic action, since all atoms, molecules, and particles are of photons. Remember, quantum mechanics determines that all particles are also waves. And, the UOE states that all waves (EME) have photon mass. The number of photons in a particle determines its mass, and its frequency.

The actions of the nucleus can be calculated as though it is constructed of separate protons and neutrons. When a nucleus absorbs or emits a "particle," it is a discrete, calculable, and a detectable amount of mass.

Also, when a nucleus absorbs or emits a "particle," the mass, therefore the intrinsic frequency of this nucleus changes.

Large particles such as the sun, earth, etc. are made of smaller particles, such as atoms. Atoms are made of even smaller particles such as electrons, neutrons, and protons. Neutrons, protons, and electrons are made of photons. What are photons made of? A photon is mass created through the lack of energy velocity. This is a true conversion of energy to mass (transversion).

Phase III
Earth and the UOE

Prefatory

As we know, our universe is a Universe of Energy. It is not void or empty as most may believe or had previously been trained to think. Space is chock full of energy. Every cubic centimeter of the universe is packed with energy of all kinds, types, and various magnitudes, in wavelengths of years per cycle up to the highest frequencies not yet discovered. These energies are, in most part, electromagnetic energy (EME) in form, and constantly moving through space at, or greater than the speed of light. All objects in space are encased in this ever moving UOE, like objects in the sea which are affected and influenced by the enveloping sea pressure. However, the UOE is not the "elastic aether" concept, unless one believes the aether is made up of super small particles moving at superluminal velocities.

We began with the small and are now starting to move toward the large. As we have learned in Phase I, EME exerts a force.

The Earth did not start as a whole planet. Nor did other planets or stars start full size. A few astronomical particles such as comets, asteroids, etc. may have originated by a division of other larger bodies, but these larger bodies were not always large.

Nothing in nature, at least not obvious, has begun its life or existence as a large or full size body or particle. Particles and other bodies in nature usually begin by the joining of two or more smaller bodies or particles, and then increase or grow from this joining (accretion). The smaller particles or bodies may often start as energy which has transverted to mass, thus causing as association with other mass. If it were true that the universe really began with a bang, it would have been only a little tiny bang, then the mass grew as it absorbed the energy in the universe.

The UOE that fills space surrounds the earth, exerting an energy "pressure" which attempts to compress the earth and all aboard. This energy is not stopped by the earth, but a small part is "absorbed" as it passes through the earth.

See Figure III.I

When part of this energy is "absorbed" by the earth's mass (as you would absorb light to create a shadow), it is converted to other types of energies, such as heat, light, etc. Light, of course, affects us little from inside the earth, but heat is a major factor in building the shape of the earth's surface through shifting "tectonic plates" and volcanoes.

Figure III.1

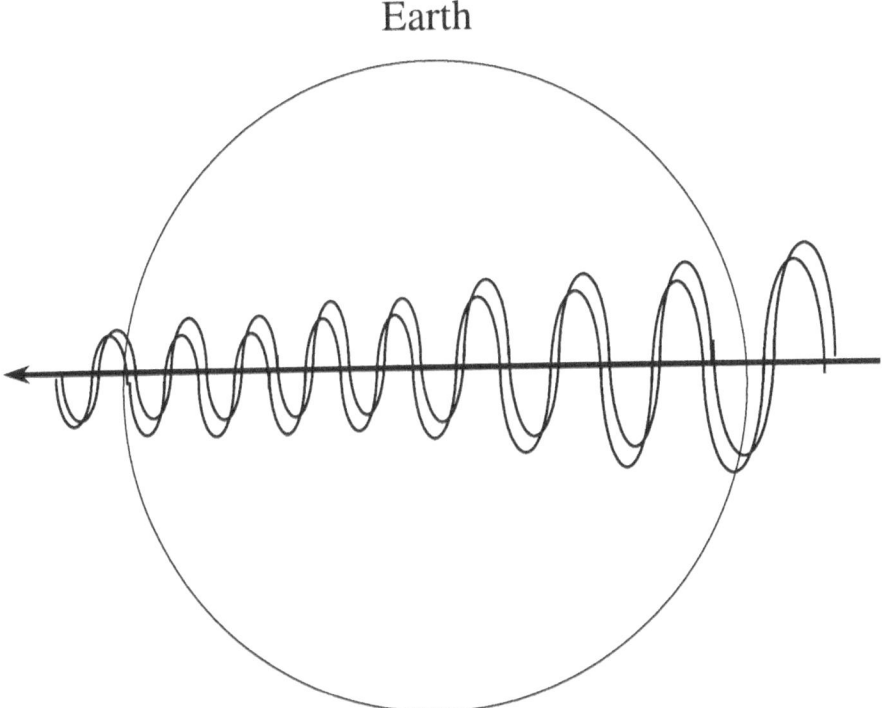

Earth

This figure shows only one "line" of energy entering, passing through, and partially being absorbed by the earth and all its inhabitants. In reality there are an infinite number of "lines" of energy penetrating the earth from all directions. In this illustration the amount of absorption is greatly exaggerated for explanation, and to enhance visual notice.

Earth has not cooled down in the billions of years it has existed because it is supplied by an infinite source of energy, the UOE. The inner two thirds of the earth is the hottest, but the highest heat concentration is within the inner one third of earth where most of the energies converge.

Objects of all kinds tend to repel other objects. If not for the UOE, when two objects come together they would repel each other, not attract. "Matter" (or mass) tends to repel other matter, not pull together. It's the UOE that forces objects together.

If gravity came from each individual particle in space, spread out over the universe, attracting other particles, then a drop of liquid in space (micro gravity) would be pulled apart by these particles. In reality, a drop of liquid is not pulled apart, rather it is pushed and held together by the UOE. Of course interferences may cause this drop to splatter, but then each part would be held together by the UOE, each drop would also become a sphere.

Earth and Moon Model

If we imagine a scaled down model of the earth and moon as two magnetic balls, one about 1 foot in diameter, representing the earth, and the other about 3.3 inches in diameter, representing the moon, they would be set about thirty feet apart. The small ball is circling (orbiting) the large ball about 10 mph. Inertia is trying to put the small ball into a tangent (straight line), making it fly away from the large ball. The total magnetic field between both balls at a distance of thirty feet is minimal, and not enough to overcome the inertial energy of the moving small ball. Therefore, the small ball will fly away from the large ball. The strongest magnets made cannot maintain the small magnetic ball's orbit.

The moon is about 239,000 miles from earth and orbiting the earth at almost 2,300 mph (relative). There is a tremendous force of inertia being exerted with this speed and mass. The combined "gravities" that are supposed to come from the earth and the moon could not be, and are not, the forces maintaining the moon's orbit. The UOE is the entire force of "gravity."

Forces of Gravity

Again, for simplification and to save much math, we'll use just a few lines of energy force for an example, rather than the infinite amount that "enspheres" the earth and moon.

We'll look at it this way:

There is an UOE force we'll call "FE" (F is force, E is earth) pushing the earth towards the moon, and another UOE force we'll call "FM" (M is for moon) pushing the moon towards the earth. The earth is shielding some of the "FE" from the moon and the moon is shielding some of "FM" from the earth. Some of the force "FE" that is pushing the earth towards the moon penetrates and travels through the earth, being partially absorbed. Some of this energy is converted to other types of energies, such as molecular motion (heat). Then the force "FE" will leave the opposite side of the earth with less magnitude. We'll call this reduced force of energy "fE" (f for reduced force).

The same is happening with the UOE pushing the moon towards the earth. This moon's reduced force will be called "fM." We now have the energy forces, "FE," and "FM," that are pushing on the earth and moon, as they pass through the earth and moon. Then they become "fE" and "fM" that are the exiting forces leaving the earth and moon with less energy.

See Figures III.3 and III.4

As these forces lose energy while reacting with the earth's mass and the moon's mass, they become "fE" and "fM," continuing on at their reduced energy levels toward the moon and earth respectively, affecting them in opposite directions, but with less strength than "FE" and "FM." The differences in the level of energies (FM-fM and FE -fE) are the forces acting to maintain the moon's orbit. See Figure III.2

Figure III.2

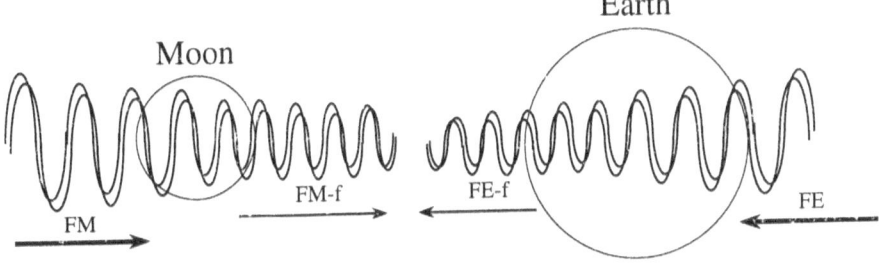

As FE and FM enter, penetrate, and pass through the earth and moon, they are partially absorbed. After they leave the earth and moon, FE and FM are designated as fE and fM to indicate they have slightly lower amplitudes. Having slightly lower amplitudes, fE and fM will have less "push" on the moon and earth in the opposite directions. Also, fE and fM will once again be partially absorbed as they penetrate and pass through the moon and the earth.

See Figure III.3

Figure III.3

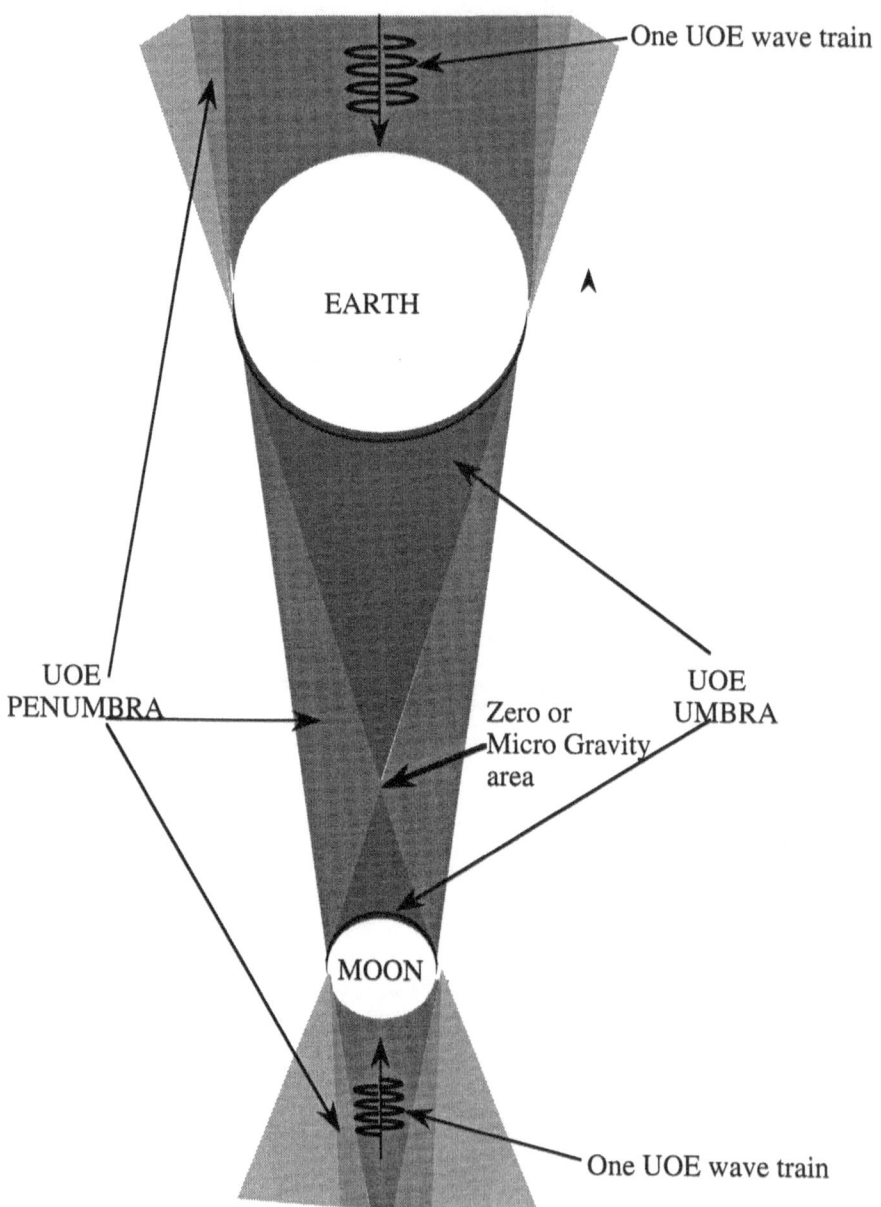

The forces are explained using only one line of energy, but of course one line of energy could not effect gravity. Space is a universe of energy, containing forces of almost infinite amplitudes and types of energies surrounding the earth, moon, sun, planets, etc. In effect, the sun, earth, etc. act as energy "shadows" for each other.

All the galaxies, planets, moons, comets, asteroids, and even atoms and atomic particles are acted upon in this manner. This is a balance of energy, mass, velocity, and force.

While we are standing on the ground absorbing a small bit of the UOE coming from above that is interacting with our mass, and the energy that is penetrating the earth on the opposite side is also acting on our mass, in the opposite direction and with less magnitude. The energy pushing us down is slightly greater than the energy that is pushing us up.

The amount of energy pressure from the UOE acting on the earth's surface is not equal at all points. It is effectively shadowed by the moon, sun, and other celestial bodies. These shadowed areas have the lowest energy pressure on the earth's surface. Objects in the moon's shadow have a slightly less force pushing them down than the un-shadowed objects, therefore these objects are pushed slightly upward by fE toward the moon. This action causes tides.

There are several general clues that point to gravity as a pressure, rather than an attraction. (1) - Electromagnetic energy (including visible light) exerts a pressure. (2) - Several notable Physicists have determined that gravity is much faster than light. (3) - The mechanics of gravity act instantly at a distance. (4) - Quantum mechanics predicts that gravity is electromagnetic energy, and it also reveals that particles are both particles and waves. (5) - Liquids in micro gravity tend to form spheres; in an "attractive" gravity this cannot happen.

Phase IV
From Nothing To Nova

Life of A Star ✪▷✪

Envision two photons, or two particles, or a photon and a particle, traveling random paths that will bring them together, close enough to shadow each other from the UOE. Consequently, the UOE tends to push them together, trapping them into a unit. This unit we will name "Tu." Tu, in time, will absorb other energies that also travel through it. This action continues pyramiding for trillions and trillions of years, subsequently Tu will continue to expand as it collects energy and particles, but is yet contained by the UOE. Then in one picosecond when Tu absorbs the one or more photons that will allow it to sustain more energy than the UOE can withstand or contain, Tu will release part of, or all of, its energy/particle content with one big nova or supernova. Sun flares are partial novas.

The type of particle that may be formed is dependent on the forms of energies or particles that combine. Planets, moons, asteroids, comets, and stars all can be created by the UOE and two single photons or particles.

Much of the released energies from a nova or supernova, having less force than the UOE, may once again combine, compress, expand and nova again—and again for eternity. This action/reaction is the balance of the UOE and is recycling nature's existence.

Particle Expansion

The physical size of any particle (including earth) can decrease while the mass increases, or vice versa. This action is mainly due to atoms transforming from lower density elements to higher density elements, or vice versa. The earth's size varies, but over a few million years its average size will increase. Generally, but not necessarily, as a particles size increases its overall density tends to decrease. This size/density ratio will also be affected by a particle gaining other particles such as meteorites, etc. The density of a particle may increase or decrease while its volume may follow directly or inversely. A particle's volume is also determined by the interactions between the atoms in the particle, and effects of the UOE throughout the particle.

The earth is still enlarging, and increasing in heat. Contributing to this growth is the vegetation absorbing light and other energy. Eventually this vegetation dies, feeds other vegetation or becomes "muck." Much of the energy reaching earth transfers from energy to plant, plant life, to "muck," and sometimes to charcoal and other natural materials, ultimately contributing to the buildup of the earth's surface over billions of years. Other contributions to the earth's mass are meteorites and the vast amount of particles and energies continually bombarding the earth. A part of this vast amount of energy is absorbed and transformed into other types of energy (such as heat) and eventually into mass.

Earth's heat is mostly generated within the inner two thirds of earth. This generation of heat forces mass to the earth's surface through volcanic action as new land mass, ocean bottom, or mountains. Also, this growing earth produces earthquakes. See Figure IV.1

Figure IV.1

EARTH'S CENTER
compressed plasma at about 50 million lbs/in^2
and about than 400,000 °F

The earth's center receives the greatest amount of converged energy, the center section absorbs and converts part of this energy into the greatest heat intensity area inside the earth. The center section is greater than 400,000 °F. The temperature decreases toward the surface at about an average of around 100-120 °F/mile. The estimated heat loss from the earth is generally accepted to be about $1.0E^{30}$ erg per century. Our calculated heat gain of earth from the UOE is about $3.74E^{50}$ erg per century (about $3.74E^{20}$ erg per century greater than the heat loss).

The earth's average temperature gradient increase from the surface toward the center is about 100-120 °F per mile. At 32 miles (~3200 °F) inside the earth, iron would soften and begin to melt. At 60 miles (~6000 °F) inside the earth, iron would be boiling. At greater than 60 miles deep, iron starts to becomes a gas. At greater than 1,000 miles into the earth (>100,000 °F) iron would be somewhat near a plasma, with all its weird properties. This "plasma" is compressed to a solid by about 40 million to 50 million pounds per square inch pressure at the earth's center, creating the illusion of a solid iron core.

Energies absorbed from the UOE contribute, not only the earth and other particles increasing in mass, but also to new particles being originated, as explained briefly in "Life of a Star." Any combination of two photons or particles may "unite" to begin a new atom. This new atom will continue to "grow" as energy, mass and other particles become "associated" with it.

The growth of a particle, and/or the earth can be calculated using a "probability in growth" (PIG) formula for the expansion of atoms, molecules, cells, particles, etc.

Time in seconds for a particle to increase to a specific size from the first two photons joining to begin can be calculated by the PIG formula:

$$t_{sec} = \sqrt{(mc/d)} \qquad\qquad [IV-1]$$

where:

    t ... is astro particle age in seconds
    m ... is mass of astro particle
    d ... is particle density
    c ... is standard light velocity ($2.99792458E^{10}$ cm/sec)

This is somewhat a naive statement, for it does not consider all the probabilities of outside influences from other particles, buildup from meteorites, and loss of mass, but these factors may be included as an association probability factor. Although, even if we use an association factor of 99 percent (99%) which contributes to growth from these causes, the results from calculations would not be near the accepted present ages of our solar system's "astro particles," including earth.

Using the PIG formula (including the 99% factor) to calculate the earth's age, we will find that its age is dated from the first particles joining is much older than realized. Science believes the earth was created at its present size. This is not true.

Utilizing the PIG formula ([IV-1]) we calculate the earth's age as $5.71E^{18}$ sec. ($1.81E^{11}$ years or 181,000,000,000 years). This age is from the first atoms combining. This number appears to be almost half as large as our tax spending.

At about 30 to 50 billion years ago the earth was about one half its present mass and was completely covered with an ocean of water and other chemical ice that had an average depth of about 5 to 10 miles. Just following this time, land masses began to peek through the ocean surface to start what are now the separate continents, creating the divided oceans.

The rising land masses are due to the earth's expansion, thus thinning (shallowing) the oceans. Also, at this time the earth was continuing to increase in temperature due to the absorption of the UOE. The earth is still absorbing energy from the UOE.

Only a few hundred million years ago (about 400,000,000 yrs.) the earth was a little bit smaller than it is now. So, large creatures may not have had it as weighty as we have now. They were fighting gravity that was slightly less than our present gravity.

About $8.3E^{10}$ years from now earth will be close to twice its present size (if we don't deter natural occurrences by blowing ourselves up).

Listed below are estimated ages of a few other solar system particles:

Sun:	$2.06E^{14}$ years	Jupiter:	$6.56E^{12}$ years
Earth's moon:	$2.57E^{10}$ years	Saturn:	$4.94E^{12}$ years
Mercury:	$4.28E^{10}$ years	Uranus:	$1.42E^{12}$ years
Venus:	$1.67E^{11}$ years	Neptune:	$1.33E^{12}$ years
Mars:	$7.00E^{11}$ years	Pluto:	$2.03E^{10}$ years

These calculations do not include any factor for outside contributions, other than EME of the UOE.

The earth is warming, but it has nothing to do with people or the environment. Most stars and planets are increasing in temperature and size. The average temperature (°F) increase per year at the surface is calculated by: °F increase/year = center temp. x $1.00E^{4}$/age (years) x density. The average temperature (°F) increase per year at the center is calculated by: °F increase/year = surface temp. change x radius (miles) x °F gradient/mile. The earth's average surface temperature increases about 0.5 °F through every 100 years.

Mass Generation Within "Our" Space

Energy is transformed into mass at the rate of about $1.37E^{-21}$ g/sec/cm^3 (or $5.71E^{-6}$ g/sec/mi^3) in our universe, and within all our "heavenly" particles. If we sample a sphere where earth is the center and the distance to the sun is the radius, we would have a sphere with a volume of $1.4E^{40}$ cm³ or $3.4E^{24}$ mi³. If the energy transformation rate ($5.71E^{-6}$ g/sec/mi³) is multiplied by the volume of $3.4E^{24}$ mi³ the total rate of energy transfer is our sphere will be:

$$5.71E^{-6} \text{ g/sec/mi}^3 \times 3.4E^{24} \text{ mi}^3 = 1.9E^{19} \text{ g/sec/sphere.} \qquad \text{[IV-2]}$$

This is equivalent to the mass of one earth created about every 10 years. To expand this sampling area, we will consider a spherical area with a radius of 14 billion light years ($1.4E^{10}$ Ly) in all directions from us. This should give us an area of $9.73E^{84}$ cm3 or $2.335E^{69}$ mi³. This is our known universe (k).

Substituting these figures into the formula IV-2 will give the total energy transformation of $1.333E^{64}$ g/k/sec. [IV-2A]

This is the equivalent to $6.7E^{30}$ suns or $6.7E^{18}$ galaxies being created per second in the "universe of our knowledge." This mass transformation is only equivalent to, but not actual stars or other particles.

We must keep in mind that there is a huge amount of mass also being converted into energy. The amount of mass converted may be equal to the amount of energy transforming into mass if there is a mass/energy relationship, and there is, "$E = mc^2$." This conjecture does not include all EME velocities.

Micronuclear Explosions (MNX)

Micronuclear explosions (or micronovas) are created in the same manner that meganuclear explosions are (novas or super novas), except that the particle does not build up over billions of years, only for seconds or microseconds, hours, or a few years, involving only a few combined photons or particles.

All objects on earth, and in space, are subject to micronuclear explosions.

These micronovas contribute much to natural radiations, and you may consider that MNX are the cause of many natural radioactive elements. Micronuclear explosions contribute up to about 90% of natural radiation from the earth and earth objects. Below is a partial list of nova cause and effects:

Supernova .... complete star explosion
nova .... incomplete (partial) star explosion
partial nova .... sun flare
mininova.................... fusion
micronoiva................. fission
piconova .... unstable nuclei

These novas can be broken down further into energy levels. A sun flare may extend a few thousand miles or even millions of miles out from the sun, depending on the energy reactions.

Phase V
Astro Energy Levels

Planet Orbital Anomalies

There are noticeable anomalies in some of our planets' and comets' orbits. The orbits are not perfectly round, or elliptical, but their perihelion will precess slightly.

Planets affect each other through their shielding of the UOE. Even far distant particles affect the earth's orbit. To look at it simply: ⊙⊙ Each time Mercury's orbit brings it to its perihelion (nearest point to the sun), it arrives slightly earlier than the preceding orbit. This inconsistency in timing calculations are due mainly to the inaccuracy of the effect of the inverse square law when used for objects that are "near" each other. This is especially true for spheres. As spheres move closer together the inverse square calculation inaccuracy (error) becomes greater.

To maintain a reasonably accurate inverse square calculation, spheres should be separated by at least 100 times the largest sphere's diameter (precise accuracy requires at least 1000 times). The inverse square formula accuracy for spheres is much less than for flat planes' surfaces. Spheres' outer surfaces turn backwards, becoming most noticeable as spheres move closer to each other. If the center of a sphere is moved in to exactly one half (1/2) the previous distance, the outer sections of this sphere will not have moved in a proportional ratio.

See Figure V.1

To make a fairly accurate inverse square calculation for spheres "near" each other. The calculation must include the integration of each point in and on each sphere (at least of the surfaces).

The geometric configuration with respect to spheres affects the gravity calculations of the closest planets to the sun, and the comets as they approach their perihelion.

Figure V.1

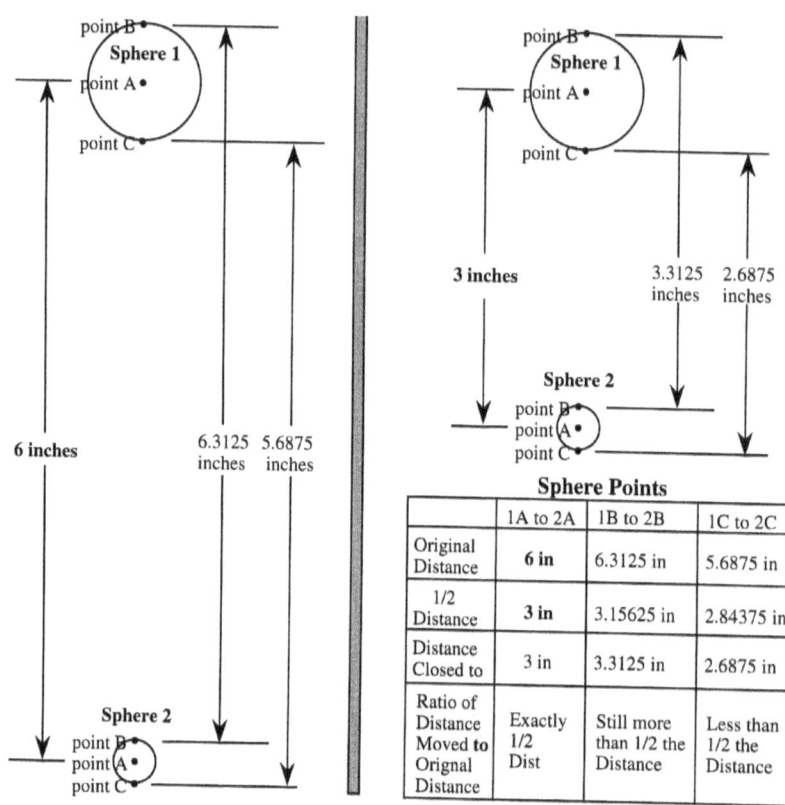

Sphere Points

	1A to 2A	1B to 2B	1C to 2C
Original Distance	**6 in**	6.3125 in	5.6875 in
1/2 Distance	**3 in**	3.15625 in	2.84375 in
Distance Closed to	3 in	3.3125 in	2.6875 in
Ratio of Distance Moved to Orignal Distance	Exactly 1/2 Dist	Still more than 1/2 the Distance	Less than 1/2 the Distance

As two spheres are moved toward each other the areas of each sphere are not changed proportionately. If the centers of the spheres close to one half the distance the leading and trailing edges close at different percentages of distances.

Mercury (closest to the sun of all known planets) is affected most by this inaccuracy of the inverse square formula due to its own position to the sun.

Comets also contribute to this mathematical puzzle, as they too approach their perihelion, for it's the sun that has the greatest diameter in our solar system ($1.39E^{11}$ cm or 864,000 miles, about 109 times the earth's diameter). This "closeness" of particles affecting the inverse square formula is one of the reasons the outer sections of some galaxies appear to rotate near the same velocity as the inner sections. The galactic centers' mass and size ratio to distance to the outer stars are much less than a hundred times.

There is a comet known as Encke's comet (discovered in 1818 by Jean-Louis Pons) that did not conform with Newton's theory of motion. J.F. Encke attempted to explain this by proposing that this comet travels through an affecting medium whose density is inversely proportional to the square of the distance to the sun. This postulate was rejected, and a proposal of meteor interference was accepted. Thus, Science again discarded another route to truth.

Our well known, and widely mispronounced name, Halley's comet (Hawley's, Hailey's, Haley's, or to some even Holy's) stands along side Encke's comet for perplexity of mathematical prediction of arrival. Back in 1910, calculations were off by over 2 days. This discrepancy was not all in the math, but also in the lack of knowledge of the UOE and the inverse square law inaccuracy for spheres.

There was a Dutch physicist who in March of 1900 submitted a paper entitled "Considerations on Gravitation" to the Amsterdam Academy of Sciences that almost hit the gravity nail on the head with the physics hammer. This gentleman physicist was Hendrik Antoon Lorentz (1853-1928). He proposed that X-rays could also produce a pressure, since it had been observed that electric waves could produce a pressure on objects (such as light onto mass). Lorentz considered that X-rays are essentially electric waves, thus he converted the action of corpuscles to these waves as a vibration type motion. He also felt that there should exist in space far more penetrating radiation than X-rays which would account for our unexplained forces that defy Newton.

Lorentz set up calculations incorporating ions. He stated that when one ion was set alone in a field of these propagating electric waves it would have equal forces all about it with no force overcoming the other to move this ion. But, the entire condition will be altered when a second ion is placed somewhere near the first. Lorentz concluded that the vibrations emitted by the second ion would cause a force on the first ion in the direction from the first to the second ion. For this to happen it was determined that electromagnetic energy must be continuously vanishing. H.A. Lorentz himself discarded this explanation of gravity. Alas! Another logical theory and a path to truth vanished.

Quasi-Orbits

The next statement presented will absolutely make no sense, and will most likely cause you to say "B.S., somebody's whacko!" But, a demonstration can be done with somewhat credible logic (even if this author is whacko) to show that the following statement is true. *"The planets do not revolve or circle around the sun; orbits are not elliptical."* At least, not the way we have been trained to perceive an orbit of one object around another. This can physically be proven by using a home made object circling a center object.

One way to create this test; is to use a variable speed hand drill with an adapter that fits in the chuck and has an extension, extending perpendicular about three to four inches from the center that holds a marking pen or pencil that can rotate around the center. When you move this marking pen along a blackboard (or whiteboard), while it is slowly rotating, the trace path will not be a circle, but a type of cycloid. The exact shape will be determined by the relative speed of the motion across the board and the rotational speed of the pen. The trace path *cannot be a circle.*

When an object (M) is standing still, relatively, and another object (m) is circling it, m will make circular or elliptical motions around object M. This entire motion is reconfigured when both of these objects are moving together in the same general direction, in the same plane. The circular orbit changes to a spiral, cycloid, or sinusoidal "orbit" or a combination of these, depending on the relative velocities and positions of M and m. Orbits are determined by the masses of m, M, the distance m to M, the relative velocities required to synchronize m to M, and the masses of the affecting nearby particles A-Z.

A cycloid trace path shape is determined by the relative velocity ratio of m to M. A trace path will be closer to a regular cycloid path when the velocity ratio is small, rather than a curtate cycloid. The trace path will be extended towards a sinusoidal shape (companion to the cycloid) as the velocity ratio increases.

We are in constant motion. This motion concerns not only the moon around us, us around the sun, but our entire galaxy is moving together in the same general direction at the velocity of over one half million miles per hour (>500,000 mph). This system motion places a different aspect on calculating "orbits." The present orbital formulae for trace paths are incomplete because they do not account for the combined and total motions of M and m objects, along with the velocities of the "fixed" stars, and the inherent inaccuracies of the inverse square calculations concerning spheres. This is partially why calculations based on ellipses do not discover reasons of preceding perihelion, etc. Due to the motion of our solar system, the planets' orbits are truly not what they appear to be.

Imagine a boat propeller (m) is "orbiting" (circling) the drive shaft (M) while the boat is tied to the dock. The trace path of the propeller is circular. If you untie the boat it will begin to move. The resulting propeller trace path is not circular, but sinusoidal and spiraling due to the rotation that is perpendicular to direction of travel of M and m.

A different trace path occurs when the rotation is on the same plane as the direction of motion. A simple analogy of this would be if you tied a small flashlight (m) to a hula hoop rim, and another flashlight (M) fixed at the exact center pointing in the same direction. When this hoop is rolled, the trace path would show a cycloid configuration by m and a straight line of light from M. See Figure V.2

Figure V.2

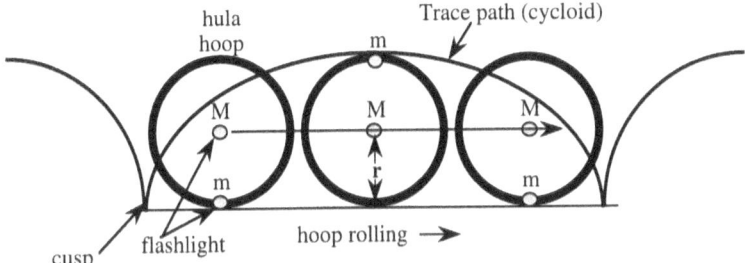

Flashlight M travels a straight path at $2\pi r$ distance.

Flashlight m travels a cycloid path at $8r$ distance.

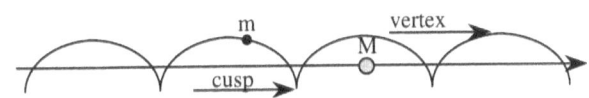

m's path is close to a cycloid when M and m velocities are

close, and M is relatively slower than m.

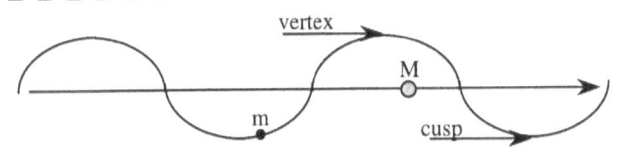

m's path moves closer to a sinusoidal path

as the velocity ratio of M and m increases.

Note: Not drawn to scale.

The apparent orbits of the planets and their moons are not true cycloidal, circular, sinusoidal, or elliptical. They are somewhat like a cycloid stretched out almost to a sinusoidal path due to the relative motions and velocities of our solar system. See Figure V.3

Figure V.3

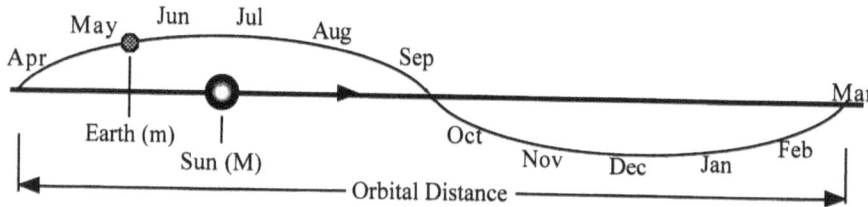

Note: Not drawn to scale.

Part (about half) of this sinusoidal path is slightly shorter than the other half, causing asymmetrical motions and velocities. These asymmetries appear and calculate as elliptical orbits; partially due to our positioning within the system.

The dynamics of the universe precludes, prevents, and excludes elliptical orbits. It is physically impossible to form a closed loop (orbit) when the center particle is also in motion.

The relative motions of our solar system particles are not apparent from our position inside the system. If one could look down at the solar system from afar (north celestial pole), the true trace paths of the apparent orbits could be recognized.

This collective motion of the solar system also causes the parcels to fall in line with the sun's path through space and the UOE. This is why the planets and moons are close to being all in the same plane. This type of configuration manifests from group motion. An analogy is to imagine a kid dragging a bunch of tin cans tied to strings down the street. The cans tend to concentrate in line with the direction of motion, however bumps, holes, and rocks interfere with their alignment, and also cause emission of energy—sound.

The earth does not orbit the sun as one would normally understand as a circular or elliptical path. The earth's path "around" the sun is the result of a combination of several geometric shapes. The earth's orbital trace path is extended into a "stretched" out cycloid becoming an almost sinusoidal path due to the sun moving in a specific direction at a relative velocity to the earth. This solar system velocity is the reason why our planets are somewhat in the same plane. As other solar systems and galaxies move along in the universe, they also will have their "followers," eventually, enter the same plane relative to direction.

The earth's actual orbital distance is at least eight (8) times the distance between the earth and the sun. The orbital trace paths may be a little flatter than depicted in Figure V.4. The arcs are slightly exaggerated for emphasis. See Figure V.4

Figure V.4

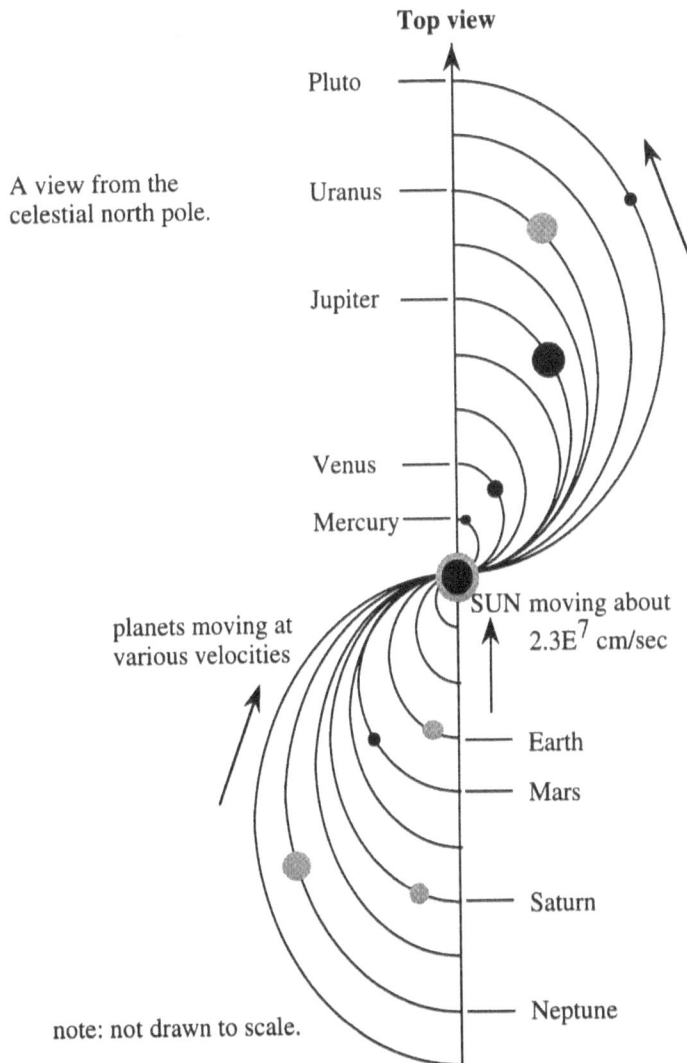

Top view

A view from the
celestial north pole.

Pluto

Uranus

Jupiter

Venus

Mercury

SUN moving about
2.3E^7 cm/sec

planets moving at
various velocities

Earth

Mars

Saturn

note: not drawn to scale.

Neptune

Our solar system is not the only one in the universe to be in motion. All galaxies, stars, and other systems are in motion, heading somewhere, at various velocities, and targets. Considering all the energy in the UOE, everything is being moved from here to there, there to here, and everywhere.

The calculations of our dynamic motions and relative velocities of the earth and the sun (and other planets, comets, etc.) present an illusion of elliptical motion.

The average apparent relative velocity between the earth and the sun is around $3.8E^6$ cm/sec (85,000 mph). The velocity for the sun (and solar system) that will establish earth's relative arc (one half of an "ellipse") is at least 650,000 mph. The apparent orbits can be calculated with elliptical formulae; but, the true shapes of orbits are closer to parabolic and sinusoidal formulae. At 650,000 mph our solar system takes over 215 million years to complete an orbit "around" our galaxy.

Inertia

Acceleration is defined as time/rate change of velocity through either speed or direction. Inertia is defined as the tendency of a body to resist acceleration. An orbit is often considered the continuous acceleration of a body toward the surface of the nucleus body, but the orbiting body velocity is enough to continuously overshoot the surface of the nucleus body.

The motion of one body orbiting another is a balance of energy, mass, and velocity. The energy is rendered by the universe of energy, and the mass is contributed by the two bodies to allow partial shielding (absorption) of the UOE to determine the velocity of the orbiting body. The continuous motion is inertia, not acceleration, even if the orbiting body is uniformly changing velocity continuously.

Acceleration alone does not determine loss or lack of inertia. Objects may be in normal orbital acceleration and have no inertial resistance. The earth accelerates as it approaches it orbital perihelion and slows as it approaches its aphelion, yet it still is in an inertial state. Lack or loss of inertia is determined by acceleration that has moved an object out of "sync" with all of its orbits.

All objects known have more than one orbit. They have many orbits. For example, the moon orbits the earth, the sun, the galaxy, and also orbits whatever our galaxy is orbiting. When an object on earth is "standing still," it is actually moving in orbit around the sun, galaxy, etc. When you throw a baseball into the air, you send it out of orbital balance with the universe (acceleration). The universe adjusts its self-balance and returns the ball to the earth and its natural orbits.

All uniform motion is curving. Any variance from the balanced curve (orbit) is acceleration. True acceleration would be in a path varying from orbit. Other than orbiting, an object cannot maintain an inertia; state and acceleration simultaneously. Acceleration must be caused by additional energy outside the equilibrium of the orbital system.

All mass in the universe is moving in an asymmetrical sinusoidal curved path. In the universe there is no inertial straight line or path. All balanced systems are moving in a curved path constructing an orbit. In our small time span we may observe only a small arc of this path. This small arc can appear as a straight line to the observer. The arc of the curve is determined by the combined foci of the orbits.

As we on earth move we are moving in a curved path. To move in a relatively straight path we require additional energy such as a rocket. But, then as all motion is relative, there is no straight path. Light and other electromagnetic propagation (EME) are also affected by mass and other EME.

Acceleration of a body is actually the orbital equilibrium unbalanced through an interfering force not associated with the orbital system. Conversely, inertia is a balanced orbital system.

On our scale, very small comparatively, we mistakenly assume straight lines, elliptical orbits, and that mass attracts mass. On a universal scale there is no straight line, mass repels mass, and orbits are mostly sinusoidal shaped as the planets move back and forth across the sun's trace path due to the motion of everything. Motion other than orbital motion is accelerated. Motion within the balanced orbit is inertia.

There is, in inertia, a balance between the plus or minus "absorptions" due to the leading and trailing sides of the moving particles, the unobstructed UOE (FE) pushing the orbiting particle toward the nucleus particle, and the partially absorbed UOE (fE) pushing the orbiting particle away from the nucleus.

For an acceleration there must be a force from one or more directions causing an unbalanced orbit. Unbalanced due to energy forces not being equal on all sides. In acceleration there will be a greater resistance to the UOE at the leading sides of the accelerating particle and all particles within it. Subsequently, there is a lack of resistance on the trailing sides. These unbalanced forces become uniform when the excess accelerating force is removed and the particle returns to an orbit. This orbit may concern the earth as the nucleus, or the sun, or another star, our galaxy, or another galaxy. Should this particle join orbits with our earth, moon, another planet, or the sun, we can observe, measure and calculate its orbit and its curve (arc).

If this particle's orbit concerns another star or galaxy millions of light years from here, we may not be able to observe its orbit or calculate its curve, but rather observe its appearance as a straight line orbit. An object on earth when put in motion may temporarily gain its inertia from a galaxy billions of light years from here. In these actions we cannot (presently) measure its curve.

A ball on a string being twirled around is not analogous to a balanced orbit of an indestructible infinitely variable velocity space ship (IIVVSS or IVS^2) orbiting earth. The ball is in acceleration due to the applied force from the twirling string. Should the string be released, the ball will attempt to attain an orbit, but it will not have the velocity to balance the force of the UOE pushing down on it. Its relative slow velocity (relative to earth's mass) does not allow the leading and trailing sides to obtain the energy absorption equilibrium. The shielding of the UOE by the earth creates an imbalance in forces, pushing on the ball and the ball returns to earth, and earth's orbit.

Keep in mind that an orbit is not circular or elliptical. Orbits are nearly sinusoidal due to the relative velocities between the orbiting particle and the nucleus particle.

As our $(IVS)^2$ launches on a path to Pluto, it leaves the shielding effect of the earth. When it shuts its engines it enters a galactic orbit to obtain inertia, no longer requiring propulsion power. Inertia obtained from its galactic orbit will move the $(IVS)^2$ close enough to have Pluto's shielding effect accelerate it to its target.

The inertial path of our $(IVS)^2$ is not a straight path, but rather curved. The arc of this curve is determined by the UOE, the $(IVS)^2$ velocity, and the shielding effect of the nucleus body. Without this balance the $(IVS)^2$ would have to power and accelerate itself to Pluto.

Inertia may be considered as the effect of an object to resist acceleration when it's standing "still." This would include objects on earth such as a rock. A rock, due to the UOE, resists acceleration, but not movement. Rocks are already moving in a balanced orbit associated with the earth. When an attempt is made to move a rock the force must unbalance the rock's orbits. As long as a force is applied to the rock, it will be in acceleration, and outside any orbit. When a force is no longer applied, the rock will return to a balanced orbit with the earth and the equilibrium of the forces acting on it. The rock was not taken from a standing still position and powered to a moving position, it was already in motion along with the earth and its determined velocity. Like the $(IVS)^2$ it is not the rock that resists acceleration, but rather the forces resisting an un-equalizing action, to move it out of a balanced orbit. Inertia is a balanced force of the UOE.

Shielded Energy

To derive the amount of energy the sun must "shield" to allow the planets to maintain their orbits, we should begin with the earth, of course. The earth as a particle has a mass (m_E) of about $5.9742E^{27}$ gram. The earth as a parcel maintains an average relative velocity (EvS) orbiting the sun of about $3.8E^6$ cm/sec, close to one ten thousandth (1/10,000) the velocity of light. To find a parcel's energy we can use the "non-relativistic" translational (NRT) formula:

$$E = 1/2 \text{ mass x velocity}^2 \qquad [\text{V-1}]$$

Non-relativistic velocity generally means less than 10% of light speed. Translational describes the combined and simultaneous movement of all the parts (or particles) within a body moving in the same (non-rotational) direction. We'll look at it this way: 🕭🕭 There is a shaft that has been tossed through the air like a spear. The speed of this shaft is less than 10% of light speed, thus it is non-relativistic. The total parts of this shaft are moving in one (and the same) direction through the air, this is translational movement. If this shaft was spinning, the integral parts would be non-translational because they would be effectively replacing themselves (their own momentum) as they spin. Therefore, when there is an object (similar to earth) moving less than the speed of light, and in a direction other than just spinning, we have a non-relativistic translational (NRT) movement.

When the values of the earth's mass and its velocity are inserted into the NRT formula:

$E = 1/2 \ m \ x \ v^2$

where:

> m is Earth's mass ($5.9742E^{27}$ g)
> V is Earth's true average orbital velocity ($3.8E^6$ cm/sec)

we get: $E = 1/2 \ x \ 5.9742E^{27} \ x \ (3.8E^6)^2 = 4.3E^{40}$ erg sec.   [V-1A]

The sun must absorb energy to allow earth orbit. But this energy level ($4.3E^{40}$ erg sec) is at the earth's center, and the total energy level the sun shields is measured at the sun's surface which is the energy we must know if we are to eventually estimate the energy in the UOE.

The earth is about $1.5E^{13}$ cm from the sun. At this distance the shielding "shadow" of the sun is only a sliver. See Figure V.5

Figure V.5

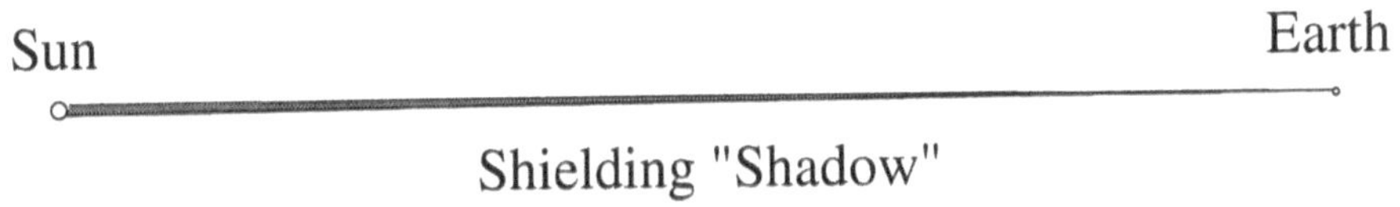

Sun   Earth

Shielding "Shadow"

Note: Not drawn to scale.

For a distance analogy between earth and the sun let's look at it a different way. If the sun was four inches in diameter (about the width of a man's hand, or the size of a softball) the earth would be about .037 inches in diameter (about the size of a spark plug gap, or 7 sheets of paper) and the distance between earth and sun would be about 36 feet. The distance between the earth and the moon would be around 1.1 inch.

A million earth volumes, or over 330,000 earth masses, would fit inside the sun. The distance from the earth to the sun is also mind boggling, when it's compared with the distance between the earth and the moon.

Using these "small" distances for comparison you can visualize the distance from the sun to the earth and the radius of our orbit. The energy shadow from the sun onto the earth would be drawn as a very thin sliver, yet it is enough to maintain our orbit (it's not a perfect orbit, but it will do for a while).

If we could set up part of this size relationship in the living room, we would visualize the vast expanse between the earth and the sun. Even more mind staggering are the distances to the sun from the far planets like Uranus, Neptune, and Pluto. Relatively, Pluto would be about 1415 feet (over a quarter mile) from our 4 inch diameter sun.

To calculate the sun's absorbed energy, we must consider the distance from the earth to the sun's surface. This distance is about $1.4597E^{13}$ cm. Total shielding effect is inversely proportional to the square of the distance (the old inverse square law trick). This inverse square effect is due to the apparent size of the circular surface area (πr^2) which increases as the square of the distance decreases. This can be demonstrated using a gamma detector and radiation, or light and a light detector. Since most folks are a little more familiar with light than nuclear radiation an explanation will be with a light meter.

If we have a round, two inch light meter, and a lit light bulb lighting an otherwise dark room. We are standing in this room holding this light meter thirty feet from the bulb. The reading on the meter is 20 marks. When the meter is moved toward the light, the meter reading will increase. As the meter gets to fifteen feet from the light, the reading is 80 marks.

How is it that the meter only moved in to half the distance but the reading increased four times? There are many explanations, but the bottom line reason is that the meter's "apparent" detector size increased inversely by the square of the distance. The formula for calculating the area of a circle is πr^2. Thus when the detector is moved in to half the distance its apparent radius doubled. As the detector apparent radius size increased by two it increased the total area of the detector by four (due to the square of this radius when calculating the area size).

If the detector is moved to a distance that would increase the meter apparent radius to three or four times, the apparent detector area would increase by nine or sixteen respectively. This geometric change occurs with any object be it round, square, oblong, or any shape you wish to make it. Round is easy to use for explaining.

To look at it in a different way: If you hold a softball at arms length and notice its apparent size, then move it in toward your eyes to half the distance, the softball will appear four times as large. You can measure this if you include squares or circles drawn on a piece of paper placed at a fixed distance behind the ball.

Now that we understand the inverse square law functions we may calculate the total energy absorbed by the sun to allow earth's orbit with the formula:

$$[E_1 \times D_1^2 = E_2 \times D_2^2] \hspace{3cm} \text{[V-2]}$$

This energy absorption of $5.93E^{66}$ erg sec by the sun is only the level needed to maintain earth's orbit. The sun also must absorb energy to maintain the orbits of all the other planets, comets, etc. The greatest amount of energy required to be absorbed by the sun to allow an orbit is the amount to maintain Jupiter's orbit. Jupiter is by far the largest planet (larger than all the other "eight" planets and their satellites combined).

All planets affect all other planet's orbits. These effects are not considered in the previous formula due to the continuous motions and changing positions of the planets. Some discrepancies in calculations are the results of each and every planet affecting every other planet, and all other heavenly bodies. Remember that these values are only WAG ballpark figures. Accurate data would require a very large modern computer to include all the variables.

The earth also absorbs energy for the moon to orbit it. This energy is equal to FE-fE. Since the moon's mass is $7.3483E^{25}$ g and its relative velocity around the earth is about $5.6E^3$ cm/sec. The NRT formula can be used to calculate the energy required to effect the moon's orbit of earth, which comes to about $1.15E^{33}$ erg sec. Using the inverse square law formula again would give us $1.02E^{36}$ erg sec at the earth's surface, but this is not the exact amount of shielding needed to hold our moon's orbit at the distance it is at. At the relatively close distance between the earth and the moon (distance to moon is only about 30 times the earth's diameter) the inverse square is very inaccurate. Also, the sun and all the other planets' shielding exert considerable influence on the moon's orbital balance.

As the earth and all its parts are absorbing energies of amplitudes around $1.0E^{12}$ erg/g/sec, then we as earth creatures also consisting of grams (some of us with too many), are absorbing about $1.0E^{12}$ erg sec. A one hundred pound person absorbs at least $4.5E^{16}$ erg/sec, or about as much energy as it takes to maintain a two ton car's velocity at almost 5,000 miles per hour.

Calculated absorbed energies are not exact due to the effects from other major planets, moons, and the other fifty plus or so major particles in our solar system. These particles contribute to not only the earth's orbit, but to each other in some major or minor way through their shielding of the UOE.

Also affecting all this inter-energy actions are all the stars and galaxies everywhere. Calculating the energy absorptions and inverse square energies of all the planets and their moons, which all effects all, would be very involved and too lengthy for this short writing. So, the tricky superfragilistic calculations will be left to the super calculating experts.

We might say we sorta' have a mutual association with the other particles in the universe. They keep our state, position and orbit and we help keep theirs.

A few formulae have been pre-calculated that we may use to find our solar system's orbital radius, distance and time. To do this a special, unique, technique familiar to many physicists, scientists, mathematicians, laymen, and professors, called the "WAG" system was applied to known values.

The definition of "WAG" is needed, so before we go much further, it means "Wild Ash Guess," or something close to that. Keep in mind that these latest WAG values are only ball park figures, and some may be way out in left field.

Using the sinusoidal orbits (true orbits), the orbit of the sun around the galaxy, the earth around the sun, and the moon around the earth can be calculated. The sinusoidal orbital distance of an object around the center object is at least eight (8) times the average distance between them. The moon orbits the earth at about 2,750 mph, the earth rotates on its axis at about 1,040 mph at the equator and orbits the sun at about 85,000 mph. Our entire solar system is moving at about 650,000 mph around our galaxy. Of course, the earth and moon velocities must be added or subtracted to or from the sun's velocity, depending on the relative directions of the earth and sun. There are no particles in our universe, or space, that is not moving (orbiting something, thus traveling in a curved path). Since particles (mass) and energy are all mutually connected ($E = mc^2$ or $E = hf$ or $\lambda c = h/\rho$). The apparent conclusion would be that energy is also not traveling in a straight line.

This brings us to a subject that is dear to, and loved by many astrophysicists and cosmologists, the big bang (BB) or the "expanding universe" theory.

If galaxies' orbits are similar to planets' orbits, then they too will appear to move away from us for awhile, a few trillion years or so, then as their orbits reverse they will appear to move toward us for a few more trillion years or so. This type action is well known by the study of our own planets as they move close then far, but in much less time than galaxies. Even the classic space travel calculations are designed for a spaceship to meet the target planets, like Mars or Venus, at their nearest point to earth.

If all we could see was a small percentage of time for the planets orbits, we would witness only a one way direction, not realizing there are two. This is one of the situations we are possibly seeing with the galaxies' orbits.

The big bang is not logical or natural. It does not follow the natural conservation laws, although many scientists are attempting to rationalize this phenomenal law breaking action. If the BB happened, the laws of nature we now understand are all wrong. Of course, some ingenious Physicists can claim the BB created our present laws of nature, but if it did it was also all wrong. All the phenomena associated with the BB can also be attributed to many different theories, but the BB theory will falsely claim them. Why has Science gone down an illogical road, in almost all fields? The BB is really a political science project. The BB believers must have bb's in their heads.

The dynamics of the universe cannot be understood until gravity and the true orbits of planets are understood.

Variations of Gravity

Gravity (UOE) is not a constant, it varies slightly. The variations are due to many actions. Actions from particles nearby, rotation of particles, varying densities within particles, and particle velocities, along with the UOE and other continuously infinite happenings in the universe such as novas, etc. which cause these variations in gravity (some may mistakenly claim these as a gravity waves caused by the big bang). Variations and anomalies of gravity are due to the continuous, very slight, fluctuations of energies in the UOE.

There are zillions upon zillions of different types and amplitudes of energies in our universe. Some of the variations in amplitudes are results of planets, stars, galaxies, etc. absorbing a small bit of this energy as it passes through them. Some of these variations may reach earth and cause slight gravity variations. The largest portion of these variations are undetectable because they occur at speeds and frequencies much greater than light, but a few variations may be detectable and even measurable. The interaction of energy depends on the type and intensity of energy, and the kind of media it penetrates. Even the layers of rock densities vary throughout the earth's interior. Solid rock may absorb more (or less) of a specific type energy than water or oil will absorb. One major variation in gravity causes tides.

If scientists could tune their "gravity wave" detectors to the high (very, very high) cosmic frequencies, they may begin to detect slight gravity variations. Sinusoidal gravity variations (SGV) are due to the earth's rotation and revolution (not the same as orbital paths). As the earth rotates while in orbit, it presents a leading side and a tailing side to its orbital direction around the sun. Gravity on the leading side is increased due to the "compression" of the UOE at about one ten thousandth (1/10,000th) the velocity of light (c) as the earth is traveling into the part of the UOE that is moving parallel to its orbit, but in the opposite direction. This compression is not yet measurable (and is really negligible).

There are other variations in gravity due to the variations in the values of the UOE. These variations are, at least, at the speed of light or greater, and are still not yet detectable with our present technology, although they should be somewhat calculable, using quantum mechanics methodology.

Of course, all these calculations may be overshadowed if we are on the trailing-edge side of our galaxy. The velocity of 650,000 mph in an opposite direction cancels the direction of earth orbit compressing the UOE. To obtain the exact change in mass one must know all directions and velocities we are moving, including the direction and velocity of our galaxy and our galaxy group. Another point is that accurate figures need powerful, accurate computers.

Phase VI
Rambling

Rambling

If the energy of one cubic centimeter of the UOE could be transformed into mass at 100%, it would create a particle over one trillion times as dense as the earth. Pure matter would have no energy and no movement. It would have no motion with respect to the universe. In other words, it would be truly stopped in space as we pass it by. Remember, energy is also a mathematical concept describing the action and characteristic of mass.

As light goes from a fast medium to a slower medium there will be "buildup" of energy at the transformation of velocity. The percent of light reflected depends on the build up of this energy in the medium. The density of the medium determines the amount of energy build up. The "reflection" or "re-emission" of energy is determined by a factor which will always be less than one (1) (representing the thickness, and density of the medium) divided by 2π.

Reflection of light is light repelling itself. When light energy is too high and too abundant for an object to completely absorb the energy arriving, the light energy being absorbed repels the oncoming energy. The absorption of light in most objects is continual. As light is absorbed it is partially transformed into heat, which is dissipated by the surrounding media, allowing for the continuation of light absorption. Therefore, if an object could dissipate heat rapidly enough, it could absorb most of the light energy bombarding it, rather than reflect it.

Can there be heat without light? Can there be light without heat? Sometimes there may not be enough to measure. Other energies create heat and light. Mass is also created by energy, slowed by mass. LRE are also minute photons (sometimes cycling in and out of light velocity). Any LRE moving in any direction, at any velocity, in any media induce wave functions.

Space is not absolute zero degrees. The interactions of the photons, from energies of $1.2E^{67}$ erg/sec/cm^2 contributes to a minimum heat energy generation by the on and off transversion of mass. As we know, the UOE consists of many energies, most of which we cannot yet detect (much less measure). Heat is an indicator of this energy, thus it is also abundant in space. The entire amount of $1.2E^{67}$ erg/sec/cm^2 is not all heat, heat is only a very small portion.

Atoms are not themselves solid, even the particles that atoms are constructed of are not solid. Atomic (and subatomic) particles themselves are constructed of as many as are the number of atoms in the earth. The atom's nucleus is not distinct separate particles. Atomic nuclei are basically plasma spheres. As a nucleus is split its mass will transform into other plasma spheres. It will also lose part of its mass as EME.

For an analogy:

A glob of jelly is a contained mass, but when it is slapped with a spoon it splits into many smaller globs, each its own shape, size, mass and even spin. In space (including atomic size space) each split piece would tend to "spherate" due to the energy pressure of gravity (UOE).

It not only takes an enormous amount of energy to create and sustain life, but there also must be a perfect balance of energy, mass, and chemistry. Energy, not only converts to matter, but it sometimes transforms to life. It takes more than a little energy to sustain life, if not there would be a zillion times more life (and many of us would be a little livelier); although, if only a minuscule amount of energy is altered from a life, the life will cease. If energy and mass in any type of combination could create life then there could be an enormous amount of life in the universe. All stars, planets, and even atomic particles would be living creatures.

It takes about $1.0E^{-12}$ erg sec up to about $1.0E^{20}$ erg sec, along with food, water, and oxygen to maintain most life. The energy absorbed by living creatures, including humans (absorbing about $1.0E^{16}$ to $1.0E^{19}$ erg sec), not only helps hold us to earth as gravity, but also is a major life sustaining energy. Living organisms transfer energy to life support and back to energy, the same as food or oxygen (such as photosynthesis).

All the energies the living body emits are not detectable. Only some of the emitted energies have been discovered so far. The normal human body absorbs about $1.0E^{18}$ erg per second (about $1.0E^{12}$ erg sec/g).

The emissions of the body through by-products or as natural waste are many. The list includes:

Amniotic fluid	magnetic energy
aura	milk
bacteria/viruses	odors
blood	others we know not
digestive fluids	perspiration
ear wax	phlegm
egg	pus
electric energy	saliva
electromagnetic energy	semen
feces	skin
finger and toe nails	stones (kidney, gall, etc.)
gases (gastrointestinal)	tears
gases (lung)	teeth
hair	umbilical cord
heat	urine

As it has been stated a few times before, the UOE contains types of energies that are not yet detected, but the total energies (and effects) of the UOE are detected on earth as weight (gravity), effecting inertia, and as resistance to acceleration.

When there are complex and extremely complicated random explanations of nature they are not correct. Nature is not complex, only man's comprehension is complicated. Nature is simple, but not as simple as human intelligence, and ego.

Physics is another means (good or bad) to understand nature. To understand is to allow familiarity, thus eventually allow use. With use comes control, and with control comes misuse. To use nature is nature, but to misuse nature is mankind. As we misuse nature we destroy it, and ourselves.

Mathematicians have little interest in numbers, they attend symbols. The mathematical process is more essential than a numerical answer. To establish a statement, formula, derivative, or expression is most relevant and paramount. Mathematics is a necessary responsibility, "enhancing" life by supporting physics. Mathematics is not just a language of communications with logic, it is also a language of logic that communicates.

We make up the world, the solar system, the galaxy, and the universe.

We are made of the astros, the galaxies, and the worlds. The infinite is in both directions, the large and the small (and all are similar). As infinity reaches in all directions, then wherever you are, whatever your size, whatever your momentum, and whatever your state, you are in the center. Any direction you move you go neither in or out, up or down, nor back or forth. Go you where and when, for no one else is, and no one else has ever been. Where we have never been, we will never go again.

A wee bit of amateur poetry:

> Go us 'er where and when
> Tho' none clever fair 'ave been
> Lo, we ne're care go again

Although poetry is not a favorite of Physicists, when attempted by an amateur it is usually comical, and the blame can be put on purpose for misspellings. Please be patient, for there are only four more fatal attempts at amateur poetry.

The energy amplitudes of the universe (and parts of) determine the "photon," "electron," et al, as defined by velocity.

Let's suppose dimensions other than our known three are not side by side, but rather external and internal, by size. The next two dimensions from us are not parallel, but larger and smaller, external and internal. The larger dimension (AsDm) is about $6.6E^{54}$ times larger than us. The smaller dimension (AtDm) is about $1.5E^{-55}$ times smaller than us. These are the next two, fourth and fifth dimensions, from us. The six and seventh dimensions are $4.3E^{109}$ times larger, and the other is $2.3E^{-110}$ times smaller than us. These are considered dimensions because they are not comprehensibly observable.

Every dimension is a center dimension, for all dimensions have the same number of other dimensions on both external and internal sides, to infinity. Like a ball, a dimension has two sides, the inside and the outside. The immediate dimension smaller than us is the atomic dimension (AtDm). The larger dimension is not named, but for our purpose we will call it the astro dimension (AsDm). Any other dimensions beyond the atomic and the astro are too far (a googolplex) beyond our comprehension. The AtDm's year is about 15,000,000,000,000 (15 trillion) of their years to our one second or $4.7335E^{20}$ (473,350,000,000,000,000,000) of their years to our one year. This is the reverse between us and the AsDm. The true difference in size structures are the time frames.

The human mind is just learning to realize sizes as large as a trillion light years, but we have not begun to comprehend sizes as small as one thousandth the size of the electron, and even less can we visualize $1.5E^{-55}$ times smaller than an electron. But, these "things" exist all the way to infinity, in both directions. And even less than less do we allow ourselves to understand other dimensions $6.6E^{54}$ times larger than us. We should have microscopes reversed to observe the external dimension, as we almost can see the internal one.

Our knowledge of the universe and cosmos is exactly equal to our knowledge of the atom and quanta, simply because they are one and the same. As infinity = 0, and finite = 0, therefore infinity = finite.

The human dominance of the earth (or even the planets, or universe) for a few million years is quite finite and of little importance by time in infinity. We may self proclaim ourselves masters of the universe some day, but even this era will pass and we will be long gone as other "sub" beings may exist for quintillions of eons. Infinity will never be ours, except in death.

We humans (this can include scientists, but few politicians) often visualize nature's way in small sections. We do understand, somewhat, the causes of a tornado or hurricane, but we envision this on a small scale. The factors involved in creating nature's way are on a scale much greater than we imagine. Natural earth phenomena are connected to the universe. Hurricanes, tornadoes, earthquakes, and volcanoes seem to be huge events, but they are almost insignificant in the spectrum of nature. Each happening on earth is a cause and effect of other happenings in the universe, and vice versa.

The UOE may have only a few relevant answers for physics, but it has something much more valuable—significant questions.

The UOE explains many old questions and phenomena, but more importantly it questions many old explanations.

The UOE answers many questions, but of greater significance it, questions many answers.

The UOE more than explains many old phenomena, it "phenomenalizes" many old explanations.

There are presently many phenomenal theories that demand to be questioned. When the concept of the UOE hits home, the physicists who are quantum mechanics may become quantum engineers.

The breakdown of observations, theories, and calculations are caused by our own technical limitation, not nature.

Scientists will never understand the dynamics of the universe until they understand the dynamics of an orbit.

Quantum mechanics applied to the thoroughbreds would do no better than the local handicappers. Handicapping is an advanced method of quantum mechanics.

Concepts now explained with quantum mechanics will eventually be, one by one, forecast through advanced intellectual science. For now, quantum mechanics is instrumental in forecasting knowledge of physics, but in years to be it will surrender to a determined relation.

Should anyone believe quantum mechanics is presently on its way out, they would be ninety nine percent (99%) wrong. On the contrary, quantum physics can contribute immensely to the UOE and the Dark Side of Light theories for many years. Quantum mechanics has already predicted gravity characteristics as being electromagnetic propagation.

Quantum mechanics should be called Quantum Procedure, or Quantum Physics. It has nothing to do with the honorable profession of Mechanics, especially the automobile racing mechanic.

Gravity is a transfer of kinetic energy into ephemeral LRE. LRE can be measured in mass, pressure, force, or motion.

A "graviton" is only a photon (m_λ') of the UOE, and may last for only $7.0E^{-29}$ sec up to $1.0E^{-11}$ sec for specific wavelengths of the UOE and light, respectively.

If energy can be slowed, accelerated, weakened, strengthened, modified, and thus force may also be altered. A force can be created or eliminated, but energy cannot. If energy cannot be created or destroyed, then matter cannot be created or destroyed, for mass and energy are one and the same (so sayeth velocity). Force is the action (and reaction) of energy.

Electrons, photons, et al, cannot do magic tricks. They consist of the same construction as the rest of the universe, they are made of natural matter. They must comply with nature's laws. Nature's laws are not truly understood. Why would scientists believe an atom is dissimilar to a solar system? The same energies (and forces) affect the astro and atomic systems alike. Nature is both large and small, and it expresses little difference in size by its laws. Why should nature vary it laws because of size? All sizes are insignificant as compared to infinity. Nature is nature, for size is finite and infinite. The actions and reactions for the astro and the atomic systems are one and the same. Whatever universal systems affect the earth must also affect the atoms in the earth.

To affect the atom is to affect the earth. To affect the earth is to affect the universe, for the earth and the universe are of atoms. As the exact same forces affect the astro and the atomic systems, then why should they not result in the same construct and aggregate?

The roads of physics we've traveled down may not have miles paced accurately, but our direction is somewhat aimed at the right destinations, holding breakthroughs, failures, excitements, answers, and more questions.

Weather

The UOE has tremendous influence on the earth's atmosphere, subsequently it contributes much to earth's weather. The UOE and the earth's rotation are largely responsible for the spin direction of high and low pressure areas.

We control (partially) the rivers, the forests, our breathing air, the lands, the atoms, and we have performed perfect disasters with all of them. Shall we control the weather, the volcanoes, or the earthquakes, the most powerful forces by earthly nature, as we do, we so perish. Just because we still have the future left, does not mean the past has been right. It is a very lucky thing for us that we cannot predict the weather with any accuracy, at any distant future. Once we can predict it then not far behind we will control it, once we control something we destroy it, or use it to destroy something else. Yes we are very lucky that weather is one of the few remaining natural earthly phenomena we cannot control (I hope).

Direction of Force

Common words "push" and "pull" are not much to think about. Or, is one real and the other not? Pull is a word for an action and a direction. Push is a word for a force. You may ask how can one be an action and the other a force when they are both opposite, but equal. There is no such force as "pull" in physics. To express another example: Jim has a rope around Tom's waist and he is moving Tom toward him. This is a direction which we can call "pull." In reality the force is "push." Why? Jim's fingers around the rope are pushing against the molecules of the rope, and the rope at Tom's backside is actually pushing him, thus Tom is being pushed by the force of the rope behind his back, in a pull direction.

You know what "pull over" by a cop means—an action, and a ticket. All forces are a result of a "push" not "pull." This action includes mass and energy, gravity and magnetism, astro and atomic (so sayeth the Universe of Energy). We will not attack magnetism in this writing.

Phenomenon of Mass (LRE)

"Mass" is a phenomenon created by specific arrays and arrangements of latent reactive energies (LRE) associated with, and affected by, the binding force of the UOE. "Matter" seems solid and real to our senses, but it is truly not solid, it is mostly space, like the universe.

As we know, if atomic particles could be broken down into smaller and smaller parts, there would be no "mass," just energy. Neither our senses, or any machine we have now, can break down the atom or an atomic particle into energy for close observance, so we consider atomic particles as mass, because we can observe it, or its actions.

Four Forces

Science claims there are four forces; a strong force (1) that holds atomic nuclei together, a weak force (2) that "pulls" atomic nuclei apart, an electromagnetic force (3) that is comprised of energies such as light, radio waves, etc., and gravity (4). Of course, the fifth force is the Air Force. These four forces will always be with us. The UOE, which holds all nuclei together, is the "strong" force (1). The "weak" force (2) is comprised of latent reactive energies. The major part of the UOE is comprised of an almost unlimited amount and types of "electromagnetic" energies (3), most are not yet individually detected. The remainder of the UOE (4) is comprised of particles such as neutrinos, mesons, pions, etc., as created by the UOE. Although, binding energy, part of the weak force, the strong force, gravity, electromagnetic force, and the UOE are all the same force (a uniforce), they are not all the same type EME (different frequencies and wavelengths).

Red Shift

The "Red Shift" theory allows the motion and velocity of an EME emitter, or receiver to affect the frequency of the EME being emitted, due to the Doppler effect. A value of frequency change is calculated by the ratio and the relative direction of the emitter.

The general accepted formula for this calculation is:

$$fo = fe/(1 +/- (Ve/c) \cos \theta) \qquad \text{[VI-1]}$$

Where:

fo is frequency observed
fe is frequency emitted
Ve is velocity of emitter
c is velocity of light (~ $3.0E^{10}$ cm/sec)
θ is the angle of the observer with respect to direction of emitter
+ is used if the emitter is moving away from the observer (opening)
- is used if the emitter is moving toward the observer (closing)

Where:

fe is $5.0E^{14}$ Hz   Ve is $1.5E^{10}$ cm/sec (closing) c is $3.0E^{10}$ cm/sec

Thus:

$$fo = 5.0E^{14}/1 - (1.5E^{10} \text{ cm/sec}/3.0E^{10} \text{ cm/sec})$$
$$fo = 1.0E^{15} \text{ Hz.}$$

However, this formula is incorrect, even if we insert the relativity effect. If the emitter was closing toward the observer at 1/2 c, the observed frequency would be 1.5 times the emitted frequency. The velocity ratio increase shows a factor of 1.5, but in this formula the frequency factor is 2.

An explanation is simple if we allow the EME emission to be emitted as pulses per second (pps). When an emitter is closing its pulses per centimeter (ppcm) will increase due to the travel time and distance of the emitter between pulses.

The correct formula for observed frequency from an emitter traveling at 1/2 c directly toward an observer is:

$$fo = fe + (fe \times Ve/c) \text{ (neglecting Relativity)} \qquad \text{[VI-2]}$$

Where:

fo is frequency observed
fe is frequency emitted ~ $5.0E^{14}$ Hz
Ve is velocity of emitter ~ $1.5E^{10}$ cm/sec
θ is negligible

Thus:

$$fo = 5.0E^{14} \text{ Hz} + (5.0E^{14} \times 1.5E^{10} \text{ cm/sec}/3.0E^{10} \text{ cm/sec})$$
$$fo = 7.5E^{14} \text{ Hz} \qquad \text{[VI-2A]}$$

The frequency change (Df) for the Doppler effect would be:

$$Df = fo - fe \text{ or } (7.5E^{14} \text{ Hz} - 5.0E^{14} \text{ Hz})$$
$$Df = +2.5E^{14} \text{ Hz.} \qquad \text{[VI-2B]}$$

If the emitter was moving away from the observer at $1.5E^{10}$ cm/sec the observed frequency would calculate to $2.5E^{14}$ Hz. [VI-2C]

Note: The relativistic effect, second-order $(1/\sqrt{1- (Ve/c)^2})$, is not included in the above formulae VI-2.

One may say how can anyone know what the original emitted frequency is from a star or galaxy. Scientists are "fairy" sure of what elements construct stars and the EME frequencies these elements emit when heated to very high temperatures. Subsequently, scientists can compare the frequency bands observed with the assumed frequencies being emitted. Thus "red shift" can determine the general direction and velocity of an emitter. Red shift does not mean a frequency band can only be shifted toward the red side, the shift can also be in the direction to blue, when the distance between the object and observer is closing.

There are two other primary causes of apparent frequency change. One is the observer's velocity and relative direction. The other is the distance and medium the EME encounters during its journey to our eyes and instruments. The first one we'll consider is the velocity and direction of the observer. This phenomena is not caused by an emitter's Doppler effect, therefore the logic will be slightly different.

If you took a stick and ran it down a picket fence, there would be a higher frequency of whack, whack, whack than if you walked. The frequency (fence boards) and the wavelengths (distance between boards) are fixed, but as your velocity changes (walk or run), the emitted frequency changes.

The formula for observed frequency is:

$$fo = fe +/- (fe \times Vo/c)$$ [VI-2]

Where:

> fo is observed frequency
> Vo is observer velocity
> fe is frequency emitted
> c is velocity of light (~ 3.0E10 cm/sec)
> + is used if the observer and emitter distance is decreasing
> - is used if the observer and emitter distance is increasing

If an observer is traveling toward an emitter at 1/3 c ($1.0E^{10}$ cm/sec) and fe is $6.0E^{14}$ Hz, the observed frequency would be:

$$fo = fe + (fe \times Vo/c)$$ [VI-2]

Thus: $fo = 6.0E^{14}$ Hz + ($6.0E^{14}$ Hz x $1.0E^{10}$ cm/sec/$3.0E^{10}$ cm/sec)

$$fo = 8.0E^{14} \text{ Hz}$$

This action also shows an increase in frequency, but through high velocity by the observer, and it is caused by a slightly different phenomenon than the emitter Doppler effect, The observer is "running down the frequency fence with an eye stick."

The other primary cause of frequency shift is the distance and medium through which we view objects (from less than 1 light year to over 20 billion light years). The UOE in these vast spaces present massive amounts of energies. The limit of almost 15 billion light years is only due to our lack of technological ability to observe any farther out.

An educated, scientific wild guess (WAG) taken to calculate the distance (in centimeters) to a heavenly body (star or galaxy) is to square the difference between the emitted frequency and the observed frequency.

$$d = (fe-fo)^2 \qquad\qquad [VI\text{-}3]$$

Where:

> d is in centimeters
> fe is frequency emitted in Hertz (Hz)
> fo is frequency observed in Hz

An example would be: $d = (fe-fo)^2$

Where:

> $fe = 6.0E^{14}$ Hz
> $fo = 4.0E^{14}$ Hz

Thus:

$$d = (6.0E^{14} - 4.0E^{14})^2 \qquad\qquad [VI\text{-}3A]$$
$$d = 4.0E^{28} \text{ cm or about } 4.228E^{10} \text{ lyr (42.28 billion light years).}$$

This is only an example. Forty two billion light years is a little too far to see right now. If the difference between fe and fo is $1.0E^{14}$ Hz the distance would calculate to about 10.1 billion light years. The difference of fe and fo represents the energy "loss" of the EME traveling great distances through the Universe of Energy. The frequencies may be corrected for the index of refraction for our atmosphere. For example: if the wavelength of $5.0E^{-5}$ cm (green light) was corrected for air at 15°C, 76 cm Hg, it would be $5.001391E^{-5}$ cm.

The velocity of an observer can cause an apparent frequency change. The distance and medium between the emitter and observer causes energy absorption, subsequently directly changing the observed frequency. This means that violet light from a star or galaxy will reach us before the red light, assuming that they are emitted simultaneously. This can be shown easily by taking a red filtered and blue filtered (or infrared and ultraviolet) photographs of the stars and galaxies simultaneously. The positions of the stars and galaxies in the two photographs **will not match**. The lower frequencies are slower than the higher frequencies. All frequencies will "lose" some of their energy from there to here.

Misinterpretation of this frequency loss is the blame for the cuckoo fallacious big bang theory.

Light propagation can go faster than its associated energy can support, but the medium can cause the propagation of light to slow through the attenuation of its energy. This phenomena is called refraction and absorptive index k. However, we will not enter into this convoluted, complicated field of optics in this writing.

Let's now approach the relativity statement that the velocity of light is the same for all observers. If you were traveling at the rate of 1/3 c ($\sim 1.0E^{10}$ cm/sec), and someone who was standing still behind you shined a blue light at you, the relative velocity between you and the light coming from behind you would be 2/3 fe. According to the frequency change formula [VI-2], when the light arrived, you would see red light.

The formula for observed frequency [VI-2] is:

$$fo = fe +/- (fe \times Vo/c)$$

Where:

fo is frequency observed $c = 3.0E^{10}$ cm/sec
Vo = $1.0E^{10}$ cm/sec $fe = 6.0E^{14}$ Hz

Thus:

$$fo = 6.0E^{14} - (6.0E^{14} \times 1.0E^{10}/3.0E^{10})$$
$$fo = 4.0E^{14} \text{ Hz (almost infrared)}$$

The true frequency emitted is $\sim 6.0E^{14}$ Hz, but the relative velocity between you and light propagation creates the appearance of frequency change of $2.0E^{14}$ Hz. You or any other detector would assume the relative velocity of light is $\sim 3.0E^{10}$ cm/sec (c) and the true frequency emitted is $4.0E^{14}$ Hz (red).

We will send you on another trip, again at 1/3 c ($1.0E^{10}$ cm/sec), but this time the emitter is traveling directly behind you, in the same direction, at the same velocity of 1/3 c. The relative velocity between you and the emitter is zero. The relative velocity between you and the blue light coming at you would still be 2/3 c because visible light does not add its velocity to the emitter's velocity. You would also observe an apparent frequency of $4.0E^{14}$ Hz.

According to Relativity the velocity of light is constant, no matter what the velocity of the emitter. The emitter's velocity does not change the velocity of light (c), only the apparent frequency, due to the Doppler effect. Once the light is emitted it travels only at the speed of light, it does not add or subtract its velocity to or from the emitter's velocity. The energy of visible light is not great enough to propagate light frequencies at speeds greater than c. Of course, if you did not know there was a blue light shining at you, you would not realize a frequency change. This would cause you to believe you are seeing red light at the velocity of c, rather than blue light at the relative velocity of 2/3 c.

We have used pulses to represent EME (light) emission for the Doppler effect explanation. This is a correct representation because light photons are actually emitted as pulses, not continuous waves. The basic action that allows the Doppler effect is that the emitter travels a specific distance between pulses, subsequently changing the distance between pulses. Two photons per wavelength is effectively pulsed emission.

A continuous emission would have zero time period between pulses, therefore the distance between pulses would be zero:

$$d = VT \qquad\qquad [VI-5]$$

where: d is distance   V is velocity   T is time

A continuous emission would preclude a Doppler effect. Therefore, light (EME) is pulsed emission, as explained in Phase I.

To account for Albert Einstein's theory stating that the velocity of light is the same for all observers, science had to prevaricate clockworks of time. Accordingly, time would be the only element to change if light velocity is constant, no matter what. The velocity of light is not a constant, nor a limit.

An observer, eye or instrument, is designed to receive and detect pulses per second (or microseconds) to distinguish frequencies and color, but observers are not designed to detect velocity, just frequency.

The inherent Relativity mistake is assuming constant velocity and changing frequency, instead of apparent frequency change caused by varying velocities, relative or absolute. This is also a discrepancy in empirical studies of light propagation velocities. An observer may discern the frequency, but not the velocity, unless the observer is aware of the original emitted frequency.

Why do scientists neglect and ignore some of mother nature's most obvious clues? Has all common sense and logic been discarded for modern "magic" theories? In Relativity when time is changed the change is apparent, meaning illusion (not true), as according to "Lou Costello" math.

We can measure only time that has passed. Time itself is a finite value to us, but is infinite in eternity. All future stated occasions are predictions, some way out and some fairly accurate, but only predictions. Time and velocities are continuously changing. As time exists in the universe it must be non-local. What value have our calculations if time and velocity are not intrinsic, reliable fundamentals? Even orbits would decay or wobble wildly without intrinsic time.

The energies associated with visible light create the familiar velocity of light ($\sim$ 2.99792458E^{10} cm/sec) in "vacuo." EME energies greater than about one erg ($\sim$ 1.51E^{26} Hz) are needed to support EME velocities greater than c.

For a summation of energy and frequency variations we will notice that:

1. An emitter's velocity can cause frequency shift through the Doppler effect.

2. The distance between the emitter and the observer causes "loss" of EM energy (thus frequency).

3. An observer's velocity can cause a shift in observed frequency through the "fence" effect.

   Causes of frequency variations:

   A) Energy

      a. distance

      b. medium

   B) Velocity

      a. observer/emitter

Just for information, the distance in space that will reduce violet light (7.69E^{14} Hz) to red light (3.95E^{14} Hz) is about 1.4E^{29} cm, or 1.4785E^{11} ($\sim$147.85 billion) light years.

Parallax Measuring Star Distance

As our solar system is touring through the universe (along with the rest of the stars and galaxies) at breakneck speeds, measurements and calculations used for distances to nearby stars do not correctly include these velocities and directions; consequently, they are not really accurate. The accepted parallax system for determining the distance to a star uses the angles to the star in reference to the sun's center as determined at six months intervals in earth's orbit.

These measurements can only be accurate if relative motion is not occurring. This is not the case. Our sun and entire solar system is in motion. Due to the velocity of the sun, the earth's true orbital distance is greater than the distance used for the parallax calculations.

The distance the earth travels "around" the sun is assumed to be about $2\pi r$ (where r is the distance from earth to the sun, about 93 million miles). The parallax measurements of "near by" stars are based on the diameter of this orbital distance. This measurement does not account for the motion of the sun and solar system.

Due to the motion and velocity of the sun, the earth's orbital distance is around eight or nine (8-9) times the distance from the earth to the sun (~7.44E^8 miles). Because a true orbit is not round, or elliptical, or any type of closed orbit, but a sinusoidal type orbit (open and continuous), a parallax measurement can be taken over one half (½) an orbit, or one orbit or many orbital distances. This would greatly increase the accuracy of the distance measured to stars using the parallax calculations. Parallax measurements do not truly account for the velocities and directions of the stars being measured.

There is another huge problem with parallax star measurements. The sun is moving about 650,000 miles per hour in orbit "around" the galaxy. This means that in one half of an earth year the entire solar system has traveled about 2.85 billion miles. The comparatively short distance of earth's orbit is nil for distance calculations (about 186 million miles, which is ~ 6% of the distance the sun and solar system traveled). This really begs for a typical question: why haven't our "brilliant" scientists figured this out?

Phase VII
Epitome

Time-Space Warps

We'll consider another strange and well accepted theory. How can well educated (very well educated, or politically indoctrinated) people and scientists believe in time and space warps? Especially warps due to the motions that all particles in the universe possess.

How can the velocity of one particle control and affect time, local or nonlocal? Does it reach over and slow down the clock next door for local effect? Can one particle reach across the universe to affect time there? Aren't all things connected, eventually in the universe? If true, then all particles in motion would eventually affect time in all other parts of the universe. Therefore, no action of time can be truly local, nor real. The universe is not chaotic, only our knowledge is.

If time is affected by velocity, and all particles possess velocity, then time will be affected by these particles moving at different velocities, different rates, and in different directions. Would it not cause complete chaos and epochs' turmoil if time depended on particle velocity?

How can relative velocity effect slowing of time? Even if velocity was absolute, it could not affect time. Time is not relative, but all velocity is. If true time exists, other than man made, it is absolute. Man is not intelligent enough to give time magic. If we cannot do things natural, then we definitely cannot do things super-natural. Empirical studies of velocity and time compression are ***seriously flawed***.

A clock traveling with relative velocity (within 10% of c) does not run slower because of velocity. This clock will not run slower than any other clock, anywhere, unless it is calibrated and operated by a specific light frequency. Then it will run at the rate of the observing frequency, either fast or slow. This action only affects the clock, not time.

> Overslept clocks at relative velocity,
> Run not more slowly tick tock,
> 'Cept for those in animosity,
> Done of others a quick epoch

Time, even though we wish it to slow, is too slow, for it will kill us long before it educates us.

> Time slows not for our will,
> Never is it standing still,
> Before us it won't reveal,
> Fleeting us, it will kill.

A space warp is just as hysterical as a time warp. If one wishes to address gravity as a warp, then it would be much more appropriate and fashionable to call it an energy warp. An energy warp is created as a particle absorbs a small portion of the surrounding UOE. These energy "warps" allow the slightly higher energies on opposite sides of particles to push these particles together. This warp force is called gravity. Time and space do not secure warps, people and scientists' minds experience warps.

As we are able to receive and detect higher and higher energies of the UOE, we may eventually be able to discriminate between the faster EME propagation and visible light energies. This, of course, would have great benefits in astronomy. If we could detect high velocity EME along with light from the same source, we could calculate the distance, velocity, and the approximate direction of motion of this source. We would also not be many, many thousands of years behind in the knowledge of what's going on in the universe.

If we should detect higher energies we most likely would, at first, call them gravity waves. In fact, they are gravity waves, but they are not the only gravity waves. All electromagnetic energies in the universe, as a whole, construct gravity.

Zero-Point Fusion (ZPF)

Lowering the temperature of particles excludes energy. Without opposing energies, particles can be merged. With experience and skill (and lots of math) energy can be reduced from several selected atoms to near their zero-point energy, combine them and "bingo" we have the required construction of fusion (ZPF). With real skill we could utilize the UOE to help combine and control our particles. Simple? Not so!

When immersed in physics do not truss and constrain yourself with limits, borders, and perimeters others have set. *Infinity is the boundary of dreams.*

Chaos

The order of things is not so orderly as we may wish. All things react in a random manor affected by other random antecedents. Random is chaos, but not to the point of no limits, adversely random is bound by chance, yet all effects are of a determinant.

How does chaos begin? First, something may appear to be orderly, but in truth to us it may be random. The finite limits of our measuring systems can deceive us. Let's begin with something simple. A drop of water is quite simple. Or is it?

A drop per second from a faucet can seem orderly. A drop emerges from the faucet forms a molecular strong surface, and falls when the weight overcomes the surface tension. This happens routinely drop after drop, second after second. As the faucet valve slowly erodes away, the drops increase in rate. As more and more water erodes the valve, the drop rate increases to the point that the following drop interferes with the preceding drop tension and weight, subsequently causing the drop to fall before or after order. This interference in turn relays to each following drop. From this time on drops are random until there is a steady flow. Once an order is interfered with it will in turn interfere with another order. This action can continue eternally, unless controlled. If chaos is uncontrollable, such as the universe, then random exists forever. Today was not like yesterday, nor will it be like tomorrow or any other day in eternity. Yet, there is order in the universe. If we cannot understand this order, then we will always believe in chaos.

Fifty billion years ago a meteor started its existence and its journey to earth, predetermined to become a meteorite. The effects that aimed it this way began more than fifty billion years ago when it set its sights onto earth and our presently Mr. Macdonald's freshly plowed farm fields in Arkansas. Fifty million years ago, or fifty years ago we could not calculate this meteor's trajectory. As it comes within our observations we will eventually begin to calculate its movement. Now that we can calculate its aim, it is no longer a random object. Although, when we first discovered it we assumed it was random.

Things are not random or chaotic; they are predetermined by the infinite actions in the universe. The electrons, et al, orbital or free, are controlled by the universe. Chaos is just an order we do not comprehend. Life depends on rhythm, most rhythms are sinusoidal. Unable to measure, calculate, estimate, or even guesstimate with a fifty-fifty chance, we assume things are random. What we can measure, control, or predict we call orderly. That which we cannot determine the state or position of we call random, random to us, but not nature.

Everything seems to be on a chaotic, random scale. Is this trying to tell us something? When plotted, most chaos has some sort of limits (randomly defined). The earth, moon, planets, etc. never repeat orbiting paths. However, their paths can be closely and quite accurately calculated and plotted.

Chaos can be calculated to a small degree, within our limits of understanding and knowledge. Weather is chaotic. Weather can be predicted for the brief future, but not for long term because it's affected by an infinite number of forces which often run in chaotic spirals affecting each other.

We can control some random actions temporarily, but only temporarily, and not even close to perfect. Even if we had the technology to control chaos, the control would eventually cease due to the outside chaotic interferences during eternity.

The future is chaotic because it is unpredictable. It is unpredictable because it is incalculable. It is incalculable because the total components involved are beyond our knowledge.

In all this, random is only random when our knowledge and technology are too inadequate to enumerate, predict, or determine present or future positions and states. Things are not chaotic, they just are. Chaos is really an expression of ignorance.

The great Albert Einstein was correct in his judgment of nature, the dice are not random (only our human ability is).

All in All

A true entity does not exist. The farthest star from the earth affects the smallest atom in the center of the earth and vice versa, subsequently any excitement in the universe will affect earth eventually, for it contributes to the UOE. Nuclear explosions ninety three million miles from earth cause events that define, create and support life, and induce death. Should one minor action on earth influence events each and every mile around the earth, surely significant happenings anywhere in the universe shall affect us and our future. It may not happen in our lifetime, but it will happen.

> Every occurrence anywhere
> Is connected to
> Any occurrence everywhere.

Each and every moment is linked to every other moment. Each and every event is a partial contributing cause and effect of all other events. Every action is a result of all other actions of nature's laws. The entirety is continuously and constantly changing. The energy emitted zillions of years ago is still, in one form or another, with us marking the past, masking the future. From the smallest quantum in an atom to the largest star, everything is "associated" and all affects all, eventually.

> The entirety is with us, 'till it's cast
> Masking the future, marking the past
> Of the wee quantum, to the infinite vast
> All time myths us, future and past
> Each affects all, 'till the end ever last

The Universe of Energy (or Sea of Energy, Ocean of Energy, Wind of Energy, Universe of Force, Sea of Force, Wind of Force, the Aether), or whatever it may be called is directly responsible for, and is the main contributing cause of many phenomena including weather, volcanoes, earthquakes, rotation of storms, the elusive binding energy, expansion and contraction of the universe, inertia, birth and death of heavenly bodies, and of course gravity.

Our present orbital calculations are almost correct because the calculations are based on time but not gravity's direction or origination. Deriving orbital formulae based on the UOE could be applied correctly.

If

If the Universe of Energy is more than a theory, it cannot be shot down by truth, for it is truth of what is. There are no time warps, no anti-gravity, no super strings, no twistons, no bosons, no gravitons, no electrons, etc. that cannot be defined, stratified, disaffirmed or accounted for by the UOE.

If and when the UOE is understood, and becomes the generally accepted basic reference, it will turn history, and science will lead a new path to truth.

If the UOE becomes incorporated with science, meteorology will probably take a huge step forward, along with new breakthroughs in physics, astronomy, and many other science disciplines, including subsequent and very important spin off theories. The UOE can explain many phenomena, consequently the sophistries and false legends associated with forces will vanish. But most importantly the UOE presents many new questions and challenges.

If things were permanent our thoughts would remain the same. Imagine one thought continuously. It's ecstatic to have thoughts turn, flee, and return.

If and when the UOE is accepted it will be reluctantly, requiring a tremendous and frightening direction change from old and dedicated convictions that gravity energy comes from the earth, and visible light is the swiftest of all energies.

If the Universe of Energy concepts are accepted, they should become tools for science to achieve special, new, exciting theories, if science can rethink and reject old tales of gravity and disorder. Of course, there are those who will always insist the world is "flat" and the sun and stars revolve around it.

If we should ever attain mind over matter, we shall undoubtedly undermine matter.

Perception

We cannot perceive all the energies that afford us life. Although we have sight and hearing, we are blind and deaf in perception of reality.

Are we aware of what we are doing? How can we be knowledgeable of our doing in the universe when we do not know what we are doing to ourselves? Human minds cannot comprehend their connection to infinity.

The universe is Nature's creation, the construction of time is man's. Infinity needs not time. Velocity cannot affect time unless man manipulates his rules to allow this, but this is not truth. Should man rationalize space-time-velocity as affecting each other, then man believes it is so. This type of rationale served for thousands of years as concepts of flat earth, and all heavenly bodies move around the earth.

Time was created when man learned the positions of the stars and planets, and enhanced when man discovered the sun dial. Time is man's invention, not nature's, only we mark time.

Information (info)

Information is an arrangement of non-information items. An example would be dots scattered in a "random" pattern, exhibiting no information. Dots scattered in an organized pattern exhibits information. Although, dots, random or organized, are themselves information, and are each constructed of information.

Should an observer of information not be aware or understand the observed information it would not be information (to the observer). An American observing dots scattered in a Greek pattern may be aware it is information, but he may not comprehend this information. This is where we are at present, we do not understand the information the universe is presenting to us. At best, we are misinterpreting it.

Unlike the American who knows the information is Greek, we haven't the slightest inkling of the language of the universe.

Information is exactly that—in formation! Or as "in a formation."

Pulsars

Some pulsars may not be the originator of their pulses. The vast amount of space between us and a pulsar contains much interference mass and energy. This mass can act as a pulsator creating rapid EME transients.

If you shine a light through a slow moving fan blade, the fan blade becomes a pulsator, and the effect will be that the light appears to be pulsating. Some Pulsar effects may be caused by slight variations in the UOE.

Dating Objects

Carbon dating of objects is not accurate for objects that are greater than 1,000 years old. Reactivation of elements by the UOE is occurring continuously. As objects decay radioactively they are also being reactivated by the UOE. Continuous activation may or may not reactivate the original elements within the object. The carbon dating inaccuracy is due to the probability of the carbon element being much older or much younger than the associated elements in a particular object. The activation probability for carbon and each associated element must be known in order to calculate the approximate age of any object. Elements are also being continuously created within the inner two thirds of the earth. These elements, also on occasion, are delivered to the earth's surface through volcanic action as the earth expands, this will affect dating results.

"Aliens think that carbon dating is when male and female humans go out to a movie."

Despite the twisted philosophical gap, there are only a few true discrepancies between the Universe of Energy and quantum theories. Should these discrepancies be resolved we'll have the ultimate unified force, combining and relating all theories and forces, perhaps revealing mother nature's master design.

Scientists and Physicists jumped head first into Relativity and its band wagon. They've committed to Relativity through flawed experiments and faulty empirical studies. The things which are constant are the inconsistent theories that have evolved through many millennium. Even modern theories (within a few hundred years) have consistently changed. Many, many years ago the earth was flat, and the universe moved around the earth. Orbits were round and then became elliptical. Gravity originates from the center of the earth. All these theories were wrong then, and are still wrong now. If all theories are eventually replaced by better theories, then all theories are wrong.

Most theories in physics are wrong, or partially wrong, especially many sections of Relativity and Cosmology. Although, curiously most of the mathematics are correct, or near correct. Theories with no facts are exciting. Facts with no theories are hum drum.

Gravity (EME) is the simplest, yet most complex of all the forces, for it is gravity that imparts the strong force, the weak force, the electromagnetic force, and even the Air Force.

Lies and Hogwash (a short list)

Listed below are a few of the many deceptions imputed to the American people:

big bang	light interference patterns
carbon dating	light speed limit
Einstein ($E=mc^2$)	moon's gravity
elliptical orbits	no cold fusion
evolution	ozone layer hole
expanding universe	time dilation
fossil fuel	time warp
Co2 global warming	UFO excuses

There are many other lies to Americans, and others, but they are political. In a way, the above listed lies are also political.

There are a few guidelines that should be observed when working with the UOE theory. The guidelines listed below are only the beginning of what may become an "arm long" list of guidelines.

Remember, any exception to a law redefines it as only a guideline.

SPECIAL RULES OF THE UOE

1) Mass interacts only with mass.

    (Energy does not interact directly with energy or mass.)

2) Forces exert push only.

    (There is no true force of attraction.)

3) Light velocity (c) is near the lower end of electromagnetic energy (EME) velocities.

4) Light velocity (c) is the average velocity by which energy and mass begin to change characteristics.

5) All wave propagation involves a medium.

6) Each EME wavelength generates about $7.3725E^{-48}$ g/λ, regardless of frequency, energy or velocity.

7) Each single wavelength of EME propagation generates two (2) particles $3.686E^{-48}$ g each), one at each inflection, through transversion (Tv).

8) Transverted mass per second (m_s) equals $7.3725E^{-48}$ g/λ multiplied by the associated frequency (Hz).

9) EME frequencies (Hz) decrease at the approximate rate of the square root of the distance in centimeters in space ($d = (fe - fo)^2$).

10) Gravity is electromagnetic energy (EME), it does not originate from mass.

11) Orbits are asymmetrically sinusoidal (companion to the cycloid), not elliptical.

12) A particle in orbit maintains acceleration resistance to an outside force.

Remember, the one word encompassing more than the "Universe" is "Creation."

The reason some people believe the sun and stars revolve around the earth is because they regard themselves as the center of the universe.

BOOK 2

RAMBLES

Political Rambling and Aphorisms
(An Attitude Diary)

Introduction

Somethin' ain't right! The need to cry out is here, again, although it's always here. Even though this voice is soft, I wish to add to the cries of the oppressed. Maybe enough voices together may shatter a few glasses which are feeding the thirst of greed.

Each of these sections are titled as separate sections, but they are all closely related and interwoven. Many statements or paragraphs can be logically interchanged and are truly correlated. Many statements, ideas, conceptions, and suggested corrections are repeated, some often, and several particular expressions should be repeated, pounded and pounded into Americans until Constitutional awareness is common place.

We must reverse this course Americans are traveling on, away from Freedom and Rights, before we leave our children and grandchildren with only dust from the ink and paper of our Constitution. This commentary while not the answer, but a guide to the solution that may help rescue Rights sadly degenerated.

This author is slightly bitter about American apathy, although I care very much for Americans and our freedom, which is diminishing with each unhealthy breath we take. Hard truth and ***personal opinions*** are expressed. Truth often hurts; but, usually the one who tells it gets hurt the most. The American people do not like to be hurt, physically, spiritually, or egotistically, even if it's for their own freedom.

This author also realizes that this writing is only an alarm clock ringing. Most people will slap the turnoff button, roll over and return to the deep sleep of apathy. Nothing short of war will "un-apathyze" the American people. If only one more person can foresee the enigma of our oppression, and move to a solution for our elusive accounting, then this writing will have succeeded.

We are the fattest, dumbest, laziest, and richest slaves in the world.

Within this writing there are, and should be, repeating of important information.

Constitution and Laws

When the Constitution is vapor, so are Americans.

Never again will there be such a conception of freedom. Its downfall will come by the "Elite." If the people would insist on, and standby the Constitution, the people would prevail, therefore we the people would overcome attempts to crush and bury us. We must insist now, while we have the Rights to do so, for in the near future we will not be allowed to insist on anything.

Freedom is far more important than morals, everyone has morals, but less than 1% of people in the world have freedom.

All laws, rules and regulations, Federal, State, or local that have the slightest hint of unconstitutionality should be struck down, waved, and repealed. Not enforcing the Constitutional Rights for everyone, even the "Elite" are a major part of Americans' problems. If we cannot understand this then foresee our Constitution dissolved and our Rights vaporized. As this happens, our problems solidify and increase rapidly. Think about Cuba, North Korea, Syria. Think about what is now happening to us, laws up, rights down, oppression up, freedom down, problems up, crime up. This is like a mathematical formula equaling the division of the Constitution and the subtraction of Rights. Not only are our problems increasing in weight, but new and extremely serious ones are evolving.

Imagine American people attempting to cope with the problems of the communist nations. We may face this sooner than we realize; if we continue to allow unconstitutional laws to flourish in the United States. If you cannot realize which laws are unconstitutional, make a list of one hundred laws and randomly throw five away. All the remaining are unconstitutional.

The United States is a come and go nation; laws come and Rights go.

If the communist nations continue to rise to democratic and republic forms of government they will meet the United States on its fall to fascism and police state. As fast as communism is turning to capitalism America is being buried by unconstitutional laws. All nations will eventually be police states.

In 25 years, or so, the Russian people will be freer than Americans. Americans will not have the right to own arms or to use any of our "Bill of Rights" because they will be overruled by laws supported by the Supreme Court criminals. The Castro regime will be nil compared to American Federal, State and local controls. Politicians come and go, but the number of laws only increase. We have well over four million laws in this country. Laws are not made for the People, but are made for lawyers.

We are stupid Americans! Yes, stupid. Stupid for letting freedom fade away. Stupid for letting our Rights vanish. And double stupid for letting our children's Rights perish.

To retrieve our freedom now requires only a legal and political battle, no loss of life, but in the future it may require a military battle with lives lost. These lives will probably be our grandchildren's. American freedom may be re-won by our grandchildren, withstanding any inherited stupidity, when they compare their inherited rights to the Constitution, if it still is in print.

What a law allows to be done to criminals also allows to be done to the innocent. A law does not normally or frequently affect criminals, but it often devastates the innocent. Laws are made to control people, not criminals. Laws that are "intended" to reduce crime often cause the reverse. These laws are not constitutional, that's why they fail. The more a law is unconstitutional, the more it tends to increase crime. All laws, especially the unconstitutional ones, are abused by law enforcement, which also increases the negative effects of laws.

Laws that are created to favor one group over another are always unconstitutional.

Dreams of lawmakers are often nightmares of the people.

The Constitution of the United States of America the Twelfth is the most important and greatest invention in the world, but the least cared for. It's the most beautiful song, but the least played. It is virtually useless in the hands of Americans, for they abuse and misuse it while maintaining apathy during its destruction.

Our Constitution is dying a slow choking death and no one will pat it on the back. As our "Bill of Rights" gradually falters, we also die, as we become less free. Less free is less alive. We are being methodically and unconstitutionally choked to death by laws. Laws that are for our "benefit." Any law that is slightly unconstitutional has its grip around our free throats. Until Americans learn Constitution, think Constitution, endeavor Constitution, and enjoy Constitution our freedom will continue to regress, and finally perish.

If you are receptive to the Constitution you are a patriot. If you are hostile to it you are a traitor.

The only reason in the world to make laws that oppress the Constitution and American people is control, control by the Elite.

If and when you pledge allegiance to the flag, you should repeat at least one of the first ten Amendments. Our rights are being violated in the false name of good, safety, or security.

When we give up freedom for security or safety we lose all three.

We should not allow the good to be destroyed with the evil, evil will return, feed on the good, and flourish. This is now what is being done by the Elite, assisted by lawyers, to Americans. This action is enforced by police, enjoyed by politicians, appreciated by government, and sadly approved by the Supreme Court criminal Justice traitors.

To The "United States" Government & its Agencies

1   Behold, the Lord's hand is not shortened, that it cannot save; neither his ear heavy, that it cannot hear:

2   But your iniquities have separated between you and your God, and your sins have hid his face from you, that he will not hear.

3   For your hands are defiled with blood, and your fingers with iniquity; your lips have spoken lies, your tongue hath muttered perverseness.

4   None calleth for justice, nor any pleadeth for truth: they trust in vanity, and speak lies; they conceive mischief, and bring forth iniquity.

5   They hatch cockatrice' eggs, and weave the spider's web: he that eateth of their eggs dieth, and that which is crushed breaketh out into a viper.

6   Their webs shall not become garments, neither shall they cover themselves with their works: their works are works of iniquity, and the act of violence is in their hands.

7   Their feet run to evil, and they make haste to shed innocent blood: their thoughts are thoughts of iniquity; wasting and destruction are in their paths.

8   The way of peace they know not; and there is no judgment in their goings: they have made them crooked paths: whosoever goeth therein shall not know peace.

9   Therefore, is judgment far from us, neither doth justice overtake us: we wait for light, but behold obscurity; for brightness, but we walk in darkness.

10   We grope for the wall like the blind, and we grope as if we had no eyes: we stumble at noon day as in the night; we are in desolate places as dead men.

11   We roar like bears, and mourn sore like doves: we look for judgment, but there is none; for salvation, but it is far off from us.

12   For our transgressions are multiplied before thee, and our sins testify against us: for our transgressions are with us; and as for our iniquities, we know them;

13   In transgressing and lying against the Lord, and departing away from our God, speaking oppression and revolt, conceiving and uttering from the heart words of falsehood.

14 And judgment is turned away backward, and justice standeth afar off: for truth is fallen in the street, and equity cannot enter.

15 Yea, truth faileth; and he that departeth from evil maketh himself a prey: and the Lord saw it, and it displeased him that there was no judgment.

16 And he saw that there was no man, and wondered that there was no intercessor: therefore his arm brought salvation unto him; and his righteousness, it sustained him.

17 For he put on righteousness as a breastplate, and an helmet of salvation upon his head; and he put on the garments of vengeance for clothing, and was clad with zeal as a cloak.

18 According to their deeds, accordingly he will repay, fury to his adversaries, recompense to his enemies; to the islands he will repay recompense.

19 So shall they fear the name of the Lord from the west, and his glory from the rising sun. When the enemy shall come in like a flood, the Spirit of the Lord shall lift up a standard against him.

20 And the Redeemer shall come to Zion, and unto them that turn from transgression in Jacob, saith the Lord.

21 As for me, this is my covenant with them, saith the Lord; My spirit that is upon thee, and my words which I have put in thy mouth, shall not depart out of thy mouth, nor out of the mouth of thy seed, nor out of the mouth of thy seed's seed, saith the Lord, from henceforth and forever.

Adapted from: Isaiah 59:

For whom shall we duly invest, the one who strives for Constitutional stress, or the one who cries for Constitutional best?

When Congress prefers to violate the Constitution, Amendments I and IX, by adding an Amendment, law, or bill to forbid the desecration of the flag, then we can kiss our Constitution good by. This places emotions ahead of freedom. Emotions, as strong as they occur, are petty and adverse to the logical maintenance of freedom. The violation of Amendment IX is the most effective resource for burying the United States of America and the Citizens without a shot.

Anti-Constitutional factions have been in the US since before World War I (even before the Civil War), and will continue to enjoy advantages in major American capacities and positions in business, education, and in government at all levels, including the agencies and armed forces, also these faction specifically are not only in the media they own and control the media.

A Democratic Republic is the peoples' freedom, instituted and maintained by Constitutional laws.

Any law that injures or suppresses one person to enhance another is unconstitutional, as is any law that is targeted at one group and not another.

The burden of constitutionality should fall upon the people, but also on the sponsors of the law or bill.

The Declaration of Independence can, once again, apply to the governments of many cities, counties, and states, along with Federal laws, regulations and codes. This 1776 Document has few words that do not describe the atrocities now being showered on Americans and our Constitution. We suffer ever increasing Constitutional violations and crimes without rebuff. We endure loss of freedom without recount. We ache and pain from oppression without rationale, yet we do not echo one of these moments toward our charges to right these wrongs. We are beguiled, misled and deceived into faith in laws, belief in our government, and in the media for our welfare. We accept the decline of our Rights rather than vacate our apathy.

American people have become the subjects, rather than the strength of our government.

The assuagement of the Constitution is real, especially the First Ten Amendments. Assaults on these Amendments come from all directions, attacks from all levels. The First Amendment: public schools, laws against freedom of speech, arrests for peaceable assembly, Federal Courts to turn back petitions of grievances ... The Second Amendment: gun laws of any kind ... The Third Amendment: still almost intact ... The Fourth Amendment: drug tests, seizure of property.. The Fifth Amendment: traffic laws, arrests without warrants or indictments.

The Six Amendment: accusers remain anonymous, long waits for trial ... The Seventh Amendment: trials without jury, kangaroo courts, lawyers ... The Eighth Amendment: tremendous bails, prisoners subject to rape and abuse ... The Ninth Amendment: other Amendments violating the first Ten Amendments ... The Tenth Amendment: state laws enacted that are prohibited by the Constitution ...

Even the Supreme Court violates the Constitution by not acting constitutionally on the grievances submitted to it. The Supreme Court's decisions are never based on the Constitution. Lawyers being allowed in our courts show that we do not have Constitutional courts. We are enslaved through contracts. We are deceived into complying with implied (quasi) contracts, contracts we would not knowingly agree with. Most Federal, State, and local laws are under color. Color ... meaning appears to be real, but is not. Traffic laws are under color. When you are pulled over for speeding you are arrested under color of law.

The Supreme Court should be occupied by people who will uphold the Constitution without regard to government lobby or "public" fad. They must maintain faith in the Framers. It appears We the People have been forsaken by the politicians and the government we "trust." We now enjoy a government of the government, by the government, and for the government.

We are misled by campaigns. We do not gather facts, nor do we intelligently process information.

It's quite a shame that Americans are letting the ink fade away on the second greatest piece of paper ever written by man, our Constitution of the United States of America the Twelfth.

It would be a super great nation, this America, if the people that love the flag with so much emotion should love the Constitution with equal intelligence. The reactionaries are crying to stop the desecration of the flag, but to do this we must burn the Bill of Rights. The Supreme Court decided, and stated that it was part of our freedom of speech to burn the flag.

No matter what kind of change of rule, law, bill, referendum, or Constitutional Amendment is enacted and passed it will still be unconstitutional. It is a very dangerous conception to desire to punish the guilty unconstitutionally. The guilty can be punished effectively, constitutionally and socially with pride, convictions, and attitude intact.

To further violate the Ninth Amendment is by far and incomparably the most risky and perilous action we can allow. It opens the door, the Pandora's box, for unconstitutional Amendments that would put the results of the Sixteenth Amendment to shame. The elite should muster beyond belief.

Relaxing the Constitution is transacting treason.

To forsake the Constitution in the attempt to control drugs is also very dangerous. All laws directed at criminals affect the innocent. And we shall, in the end of freedom, burden a police state, still with drugs.

Laws vanquish freedom, they do not conquer crime or corruption. Truth in education will conquer crime. Faith, and maintenance in the Constitution and education will diminish crime and enhance priceless freedom. Our inherited gift of the Constitution should never be weakened, abused, or misconstrued. For anyone who does this is an enemy of the people.

Few have died from making laws, but millions have died from making freedom, which is more valuable?

The mitigation of the Constitution is no longer being slowly, methodically directed. Our Constitution is now under a rapid and direct attack from the Elite and their supporters. It is being chopped, wacked, trimmed and gouged. Drug laws, drug testing, gun control laws, new proposed Amendments and Constitutional changes, Supreme Court decisions, Federal laws and regulations and State laws are a few of the many vehicles being utilized by greedy Elite, Communistic politicians and many other American internal enemies. These vehicles are supported by the media, and believed by Americans. These assaults are the most aggressive and dangerous ever, for they are hidden inside partial truths, disguised lies and factions scheming for changes that conceal perils for Americans' safety, security and freedom. The Elite insist these changes are for our good. Ironically, Americans believe this, and insist on them. Americans still believe they are not brainwashed, and are free.

The damage Americans do to themselves can be kept to a minimum by limiting the power of the "elected" in government. Factions well trained, educated and deeply imbedded in the political system do not have the welfare of Americans at heart, not do they design to enhance our freedom, safety or security. When our laws are more important than our Constitution, our Police State has begun. The drug problem, as abhorrent as it is, is not awesome enough to sacrifice our Constitution for, or any part of it.

The Constitution needed to remain in tact in order for crime to be conquered. To return to our Constitution we must first win the war on apathy, secondly, diminish the crime of ignorance, and thirdly, eliminate the monstrosity of political greed. There are many dangerous laws not being challenged. These laws have led us to the police state we now enjoy. When laws are derived from personal belief they initiate a police state. As a police state begins, the right to own arms is attacked. Taxes increase, for any reason. False "causes" (such as "public safety" or "global warming") initiate controlling laws. Crime increases and more laws are made, until there is a spiraling descent to an unconstitutional police society. We are now in the spiral. Politicians have not only failed the people, but they have become traitors through deceit, deception and lies. Apathetic Americans who want to keep these politicians in office have failed true Americans. We are now Constitutional failures.

The Constitution must be upheld at all costs, if not, it will cost all.

Unconstitutional laws (over 99% of all laws) do not serve the people, only the controllers of the people. When a law does not enhance freedom and support the Constitution, it is devastating to the innocent. Most laws created are to serve the greedy, the selfish, the spiteful. and the rich. All the first ten Amendments are equally important to preserve the freedom of the United States of America the twelfth, and its people.

When one Amendment is forfeited others will soon follow.

Americans are losing the freedom cherished and envied by all other nations through apathy, gullibility, and greed.

Variance from the Constitution, for any reason, is too dangerous to leave unchallenged.

In congressional legislative assemblies many changes and laws are created, most, if not all, are directed at diminishing the rights of Americans. It is the same for State and local legislative assemblies. The most important and critical Amendment, Amendment IX, the glue that holds the "Bill of Rights" together, will continue to be flagrantly violated. Covert factions have been designing to control the government, armed forces, schools, business, and the media since before the Civil War. They have not taken recess in their endeavor to quell the rights of Americans. In this endeavor, they have succeeded.

Laws, especially unconstitutional ones, increase, not decrease the number of criminals, subsequently crime increases. Our real enemies are in this country, not foreign countries. Our greatest enemies are ourselves. We are stupid indeed. As our other enemies conceive the lecherous means to preclude one Amendment, so may they extinguish the rest.

When one individual's rights are lost then millions of society's rights are lost, for millions of individuals are society.

There is no end until our entire Constitution is buried while we sit still, on our recliners, being brain-washed by believing the boob tube.

The Elite continue their onslaught by taking away the controversial Rights, such as gun Rights, and convince us it's for safety. Next goes the Right to assemble, for they accuse us of disturbing the peace, or we "may" do something "bad." Then they chip away our Rights to be secure in our persons, our home, our papers and effects, then assure us it will quell crime. The Elite compel us to testify against ourselves, persuading us to believe it will eliminate drugs, crime, and terrorists.

The Constitution is the Supreme Law of the land (Article VI, paragraph 2).

Anyone mitigating the Constitution is a traitor to Americans. The rationale that we are all subject to contracts under Admiralty is traitorous. Judges supporting this rationale are traitors.

Our government should protect us from enemies abroad, not create them within.

Our enemy is intelligent, scheming, cruel, cold blooded, and determined. The "moralistic" areas such as gun laws, drug laws, traffic laws, and alcohol laws are highly disputable as right or wrong, and are perfect tools for propaganda, control and oppression.

Remember, the factions who endeavor to conceive and enforce gun control and other unconstitutional laws are either sons of Cain, communists, crooks, power hungry, truly stupid, or maybe all of the above.

As we give up our rights little by little the loss is barely noticed. In the end we will have discovered too late that: *When we sacrifice our freedom for safety or security we soon lose all three.*

Who really benefits from laws and regulations? Think about it, we the people don't. Let's use gun control for an example, since it's highly controversial. First, who is hurt by gun control? They are manufacturers, dealers, shops, clubs, sports, and self protection (people). Second, who benefits from gun control? These are law enforcement, politicians, courts, lawyers, communists, criminals, and the Elite. This list itself explains much, and why gun control is pushed by the Elite, who control politicians, courts, law, lawyers, etc. This list is also connected to all unconstitutional laws and regulations.

A great government will father its people, not mother them, however, it seems the American people are breast fed.

Governmental agencies (e.g. FAA, FCC, IRS, FBI, NRC, CIA, and the various drug agencies) use the end to justify the means. But, the end never comes, it lingers on, and the means continue to undermine the rights and freedom of Americans. These agencies have expanded into great monsters more frightening than the Framers wildest nightmares. These agencies are like naughty children pushing momma and daddy Americans to the limit. However, it appears, apathy has no limit.

Weak Rights invite strong controls.

If you ever read the Presidential Executive Orders you would freak out from the unbelievable controls our government exerts on Americans. For instance they state Americans are the enemy of the United States (not the United States of America the twelfth). The United States is a corporation, the United States of America the twelfth is supposed to be our country, but is suppressed by the United States (US). All laws, regulations, codes, etc. are under the United States. The US flag is red, white, blue and yellow. The yellow is the fringe (tassel) on the edge of the flag. This is also the Admiralty law. To be under the United States of America the twelfth, you must be a *Preamble Citizen*, not a United States citizen.

We the people of America are lied to and bamboozled into believing that laws are for our own good (very few are).

Laws are not made for the People, laws are made for lawyers, subsequently lawyers are another major cause of unconstitutional laws being levied on Americans. One example is when a lawyer enters in politics and becomes a "lawmaker." While in this position the lawyer initiates many laws. The greater the number of laws (most unconstitutional) the more cases there will be for lawyers. Many laws, when passed, make instant criminals of innocent people. What we have done yesterday, and maybe all our lives, becomes illegal today.

Although, most laws are unconstitutional they are supported by law enforcement and courts. These lawyers, judges, and officers are also traitors to Americans.

We thought our Constitution was maintained. We "assumed" it was maintained. How stupid assumption is! Our Constitution is somewhat similar to a "Ferrari" rusting on the ocean bottom. It was built well, designed for freedom of movement, absorbing many bumps and ruts with much thought and labor involved by truly dedicated creators. It was trued to endure speed, to last and win challenges. This feat requires skillful driving backed by a select group of engineers and mechanics. Now, regretfully, the mechanics are toy makers, engineers turned into bankers, the thought and labor turned into greed, the race fans are all apathetic, and the sea is a junkyard. This precious "Ferrari" sits and rusts, resisting the possibility to be revived and reinstated as designed. Modern engineers have thrown out the true design and have conceived a plastic toy, a tin engine inside a deceitful design on wheels of money, driven by greedy quislings. Our "Ferrari" has been discarded, hit with false starts, called on black flags, and fuel rationed, with corrupt officials controlling the race. The apathetic fans care less who wins, or how fast, and even what hostilities this precious instrument may encounter. The race is fixed.

We must install politicians that serve us, rather than ones that control us.

We must install politicians that aren't politicians.

Drug testing for any reason infringes upon the people's Rights under the Constitution of the United States of America, the twelfth, Article VI, Section 3, and Amendments IV, V, VI, and IX,

Gun control laws infringe upon the people's Rights under the Constitution of the United States of America, Article VI and Amendments II, IV, V, and IX.

Freedom is like a ladder. The rungs are solid constitutional laws for our nation to stand on, and ascend to unchallenged freedom. The side rails are the First Ten Amendments, constraining, yet supporting laws that enhance our nation's ascent to the top rung of liberty. Once at the top, our nation may vary back and forth, up and down somewhat, but will remain stable and free as the rungs rails are intact. The nation's downfall begins as the strong, slip proof rungs are replaced by shoddy, cheap spokes that cannot be supported by the side rails. Then, new side rails are designed to hold weak imitation rungs. This unconstitutional ladder no longer supports a free people, but precludes an ascent back to freedom. Many carpenters' lives will be lost in the attempt to rebuild a new ladder. The controlling greedy Elite will only supply cheap material, and charge very high prices, to build this ladder. Thus, our ladder to freedom crumbles to the ground. Now we can only use step stools and cannot reach the freedom peak ever again.

Elite and Quislings

You may easily and readily realize what and how crooks deceive, but you must search with all your wisdom to understand the treacheries of quislings and apostates. Quislings are puppet apostates who attempt to control, undermine, or mitigate the constitutional rights of any American.

Freedom is esteemed and wonderful, too bad Americans wasted it.

The conception and preservation of the Constitution was paid for, dearly, by millions of young lives, not only American lives, but all who fought against the United Colonists, and the following wars.

The larger the lie, the greater the elegance.

Anyone agreeing to an automatic death penalty for someone who murders a cop, but not for someone who murders a 3 year old innocent girl, is truly brainwashed. This appears to be most Americans. Many go along with the idea of the automatic death penalty for cop killers because they don't think for themselves, they follow others, who are quislings.

The bright have made this nation free, the dim are making it bound.

A silver tongue can convince you that you're wrong, but you should keep believing yourself until you've examined the true context, instead of the flashy silver words.

Criminals may steal your property, your money, and even your life, but the Elite steal all this and your freedom, your children's freedom, and their children's freedom. While doing this they pick your pocket of wisdom and perception to prevent you from realizing the truth of the their crimes.

They who listen to anti-constitutional canards shall see their freedom soar to mars. As the world is controlled by quislings, we should dream of space travel.

If you fight against the Constitution you follow, when you fight for it you lead.

There is sort of a catch 22 in politics. Political candidates accept large donations from business to run attractive, effective campaigns, and get "elected." Their interest is in office and business, not the Constitution and Americans. When public attitude is influenced, control of the public is effectively applied. The politicians tell the people what the people want, not the other way. This is another covert oppression method. The "Black" community, above all other people, should have a tremendous interest in preserving the Constitution. However, they too have an abundant share of quislings to endure. The greatest problem is most Americans of all races and creeds are brainwashed and now apathetic to what is happening to our freedom.

Whom shall we cheer, the quislings who infringe on our freedom or the patriots that struggle against the quislings? Whom shall we trust, the ones who are destroying our Constitution or the ones who are fighting for it? Many lives have been lost fighting for freedom.

We cannot see the future unless we've listened to the past.

The American people are the most ingenious and imaginative people in the world, yet we are also the most brainwashed, apathetic, and gullible. We are wimps that do not stand up for our freedom. We fear our government.

Doesn't it tell us something when we fear our government? Our government does not fear us, but it is afraid we may learn the truth. That is why they continuously create adversity among people. Adversities such as crime, drugs, "war," accusations of each other, and many other distractions to misdirect our attention. The Elite have added another means to control people, "global warming." They will add many laws and controls with this new adversity. They will also create huge funds for themselves, such as they do with income taxes. Not one penny of this tax goes to Americans, it goes "over seas." They will create more taxes, and taxes create control. Our government runs us, not vice versa. The Elite do not take control for granted, they continue to create more control. We, the apathetic, mollycoddle Americans should wear kerosene soaked rags on our ankles to keep the ants away from our candy asses.

The Elite and their quislings not only take facts and add imagination, they also imagine facts.

Our freedoms are being mitigated by inside conspiracies guided by outside contingents. Most politicians and lawyers do not truly wish to eliminate crime, it's their bread and butter. It is our "elected" criminals that are making laws to control the people, not crime.

The most dangerous criminals are already off the streets, they are in government.

It seems when propaganda isn't true we believe it. We rejoice in our belief, rely confidence in our information, and assure ourselves of our absoluteness in our decisions. Thanksgiving day is coming for us. We've been fattened for the methodical slaughter. We've been fed partial and twisted truths, and straight lies, and we're dumb enough to accept the established media and its daily propaganda.

Half the truth is a whole lie.

The question is "who makes laws, criminals or lawyers?" Yes, You know the answer, both.

Americans hate to be told they are duped and brainwashed. Each person thinks they are there own person, and no one else controls them. How gullible, how egotistical, how self centered, how brainwashed we are. Americans follow, not lead. This is why fads are made. Leaders invent fads for followers. Maybe we are lucky that most Americans are followers, because most leaders are wanton and designing for control, yet, we desperately need leaders to protect our freedom, and contend with the Elite that are designing to destroy our freedom, or what is still left of it.

Legislators and Congress speak for many, except the free. Americans have the tendency to believe in anything, except the truth.

People that cry, moan, and whine about spending money for space exploration are the ones who spend billions of dollars on entertainment. These are also the same people who whine about using animals for medical research, yet reveal themselves as hypochondriacs, spending billions on quackery.

There is little wrong with enthusiasm in sports and entertainment, except when it carries a much higher priority than freedom. Americans revere, worship and honor people in petty tasks. Sports and entertainment are used by the Elite to distract us from the real problems.

Unproductive, vain, and superfluous occupations in the fields of acting and sports are viewed as important. Much time and wealth is wasted on devotion to fleeting pleasure while many others, unattended, are enduring unnecessary, starving deaths. Great creations, imaginations, and innovations of tremendously talented people are wasted on the lies of acting and sports. If the money and effort thrown at entertainment were instead spent on science we would have cured cancer and many other diseases by now. The wealth spent on entertainment is only exceeded by the wealth spent on vanity and taxes.

"Social Security." These two words are antonyms. They do not belong together (same as "military intelligence"). Security of any kind, especially monetary, does not evolve from any social program in a democracy (or any other government). The evidence is the nightmare monster the S.S. and welfare programs have degenerated into. Sadly, Americans not only allow the government to exploit and expatriate social requirements, but we seem to insist on it.

If anyone has slept for the last fifty years and woke up in the year 2008, they would find they no longer have the right of privacy, nor the right to defend themselves, nor the right not to testify against themselves, nor the right of freedom of speech, nor the right to have any rights. All this shall be in the design wake of the Elite, authorized by the Supreme Court, approved by apathy, and encouraged by self interest groups. The belief that our freedom is not being intentionally dissolved is our worst and most dangerous problem facing Americans. We have been brainwashed to fade away from terms as "conspiracies." We've been trained that "conspiracy" is a bad word. The conspiracy that is enslaving us is enormous. It involves world leaders, including the U.S. President, the U.S. Congress, all the major country leaders, and ninety five percent (95%) of the media, which is owned by the Elite.

We are a sorry lot, we modern folk. Our founding fathers, the framers, forwarded their gallant feats of democratic achievements, only to suffer a defeat caused by their own descendents' apathy.

Along with the framers, all the young men and women who have given their lives, limbs, and time in war and peace to gain and maintain freedom, may have sacrificed all this in vain. And, this is because of the stupidity, laziness, gullibility, and most of all apathy of modern Americans. Americans refuse to believe there is a massive, large scale, monstrous conspiracy of deception that has ever been achieved in this world (just for us). If there is one thing the Elite have working for them are silver tongues. Why don't we adopt a new national bird, the ostrich.

The Elite and self interest groups manipulate news media with half truths, whole lies, imagined facts, and big bucks.

It is easy to see why other countries dislike Americans. They realize what we don't. We are too involved with day to day problems that we are giving up our kids precious freedom. We work and save to educate our kids, but do nothing for their freedom. A very sorry attitude we Americans have evolved.

It matters little if Americans are Constitutionally educated, because we are ignorantly apathetic. Apathy is not ignorant, but stupid.

Ignorance can be cured, but stupidity is fatal. It would be inspiring if Americans were only ignorant.

Yes folks, We the People and our freedom are definitely under siege. Our rights and freedoms are being eroded by the Elite and their quislings, with the sands of deceitful politicians.

We do not know who our true enemies are until we elect them.

New reasons for new laws will continue to be created. One of these is "Global Warming." This is not the first and only time we have been deceive and lied to by Scientists. Their are many lies set on Americans, and the world. Most of these lies and deceits target the heart, rather than the mind.

A short list of lies follow:

aircraft jet contrails (chemtrails)
Amendments XI, XIV, XV, XVI, XVII, XIX, (not Constitutionally ratified).
Bible code
Microsoft
citizen (Citizen)
Constitution/amendments
crime doesn't pay—ask Rockefeller, Rothschild, Congress, etc.
da Vinci code
National deficit

elliptical orbits
evolution
expanding universe/big bang
freedom for Americans
freedom of information act
gas problem/prices/shortage
global warming
government agencies (FBI,CIA, FEMA, IRS, etc.)
government serves citizens
holocaust
home of the brave
interest rates
judges, lawyers, prosecutors, etc.
Kennedy assassination
land of the free
laws under color (federal, state, local)
light speed limit
marriage ("children")
money
moon landing
Murrah bldg. bombing
national security excuses
no brainwashing of Americans
no cold fusion
ownership (homes, cars, property, "children", etc.
taxes/IRS
time dilation
time warp
traffic "laws"
TV, radio, news paper ads
TV, radio, news paper news
UFO excuses
United Nations
United States vs. United States of America the twelfth
voting/elections
Waco holocaust
wars on drugs, terror, etc.
world Trade Center attacks

This is not an all inclusive list. Many other lies are laid on us, some we suspect and some we have no idea they are lies. Silver tongues deceive us.

When we take our freedom for granted, someone else will take it for good.

Laws that are created are for the Eleventh Amendment and up. There are no real laws made to protect the First Ten Amendments.

Drugs

Thousands of people have died from having no safety or security, but millions of people have died from having no freedom. Which is more precious?

If we have to sacrifice any part of our freedom to fight a war on drugs and terror we have already lost the war.

Has Congress gone bananas? Besides their onslaught on the First Ten Amendments they tried to initiate a law to allow Drug or Custom agents to shoot down civilian aircraft. Of course, these gung ho, trigger happy, idiot agents would not abuse this thrill.

The drug and terror problems are insignificant compared to the problems that are emerging from the oppression being committed in the futile and witless attempt to eliminate drugs, terror, and "global warming."

Foreign Nationals

Do Americans believe we won the Second World War? If we take a good look with some perception and insight, we should begin to comprehend how, who, and why Americans are not in control of America. Most major countries hold vital positions of advantage over us, in many more ways than we suspect. American laws are not on equal plateaus with foreign laws.

Foreign laws hardly afford Americans to sue them, yet foreign nations and their people can sue Americans. If our "friends" are cutting our freedom and stability, how large a gash are our enemies slashing? Europe, South America, and the African Nations, friend or foe, are all affixed to our backs like leeches. They are worse than leeches because they're sucking the sustenance from our wallets and our freedom. Americans are not free from foreign governments, and even less from our own. We have been cheated, bamboozled, brainwashed, mesmerized, and deceived by our own government about foreign nations and "wars." It's no surprise foreigners dislike Americans, we're stupid.

Americans have invested trillions of dollars in foreign nations, who in return use these funds for effecting their interests, which are not in American interests. The most avid of these are our former deadly enemies Japan and Germany, along with our present "enemies" Russia, China, Iran, Libya, Syria, Cuba, etc.

Japan and Germany could not beat us with armies, but they've conquered us with economics, thanks to our traitorous government. Some of the richest people in the world are Japanese, compliments of Americans. The Japanese economic strategy is to use American laws, economy, and innovation for the ground base foundation of economic control.

If our cowardly American politicians had a few smarts and a couple of balls, they would wave in the face of the Japanese, the documents of their Unconditional surrender. Then, command them to comply with our national laws.

Our "friends" the Japanese are only one of many major American dilemmas. Other "friendly" nations contribute to our future and downfall by flooding the United States with narcotics. In turn, we as good Americans, caretakers of the world, flood these nations with billions of dollars of our hard earned, unconstitutionally taxed, dollars.

We are on the same heading with China. Soon they will conquer us economically.

If the Chinese execute their own for just chit chat, should they, if empowered, attend mercy for us?

Our enemies, inside and outside America, have been peculating status, rank, positions of dominance for over two hundred years. Don't you think they've advanced a bit? They've now succeeded.

The major portion of our products used to be "made in Japan." Look at their rise to economic dominance throughout the world. Now, the major portion of our products are "made in China." Guess where their economic position will be in a few years?

We need the Chinese government like we need the North Korean government. Our loveable government is going to fund both these dangerous nations. Our history has shown that we are fools for fantasy.

All nations that we have funded has had their Elite mingle with our Elite to further degrade American freedom.

Guns

When gun laws require people to turn in their guns, who turns in their guns? Law abiding people, not criminals. This leaves the people unprotected, and the criminals with a great advantage, subsequently crime increases. The National Rifle Association (NRA) and other arms clubs are accused of being self interest groups. These organizations should be pointing the finger (you pick which finger) at the accusers as traitors. These organizations are defending the Constitution, but the Elite and their quislings are attacking them. The anti-gun factions are oppressing our freedom and the "Bill of Rights."

Did you know the first ten Amendments are for the people, and the rest are against the people?

It would seem, to intelligent people, that the ones defending American's freedom are the good guys. The silver tongue traitors wear the black hats, they are against the Second Amendment.

If one is against People's Rights, then one is against the People (this includes the government).

The quislings trying to ban semiautomatic weapons, are only under the order of the Elite. They claim it will take back the streets from the criminals. Banning guns at any level will hurt the people, not the criminals.

Any action taken to mitigate the Rights of the People triggers fatal injuries to the Constitution.

The act of violating the Constitution by law makers and law enforcement echoes the apathy and attitude Americans have developed. People believe that if it doesn't affect them, then they don't care. People do not realize that anything affecting the Constitution affects them directly. If law enforcement can throw a drug dealer or suspected terrorist in jail, without bail or trial, then they can do the same to anyone, all they have to do is accuse us.

When the Constitution is injured, Americans suffer the pain.

The Elite know Americans are easily brainwashed, so they use the greatest brainwashing device ever invented, the television. News media are not the only programs in the TV media to conduct brainwashing, almost all programs on TV fulfill this function. Entertainment is the major brainwashing media the Elite utilize, and we pay for it ourselves. We fund our own brainwashing, isn't that a neat turn-a-round. We pay for the degradation of all our Rights. We fund our own enslavement, by paying income tax. Americans are played for fools by the Elite and their quislings. What ever happened to our American intelligence and ingenuity? The TV happened, that's what.

The greatest, most effective forces to combat crime are the People, truth, education, and the Constitution, not laws. The British mentality of 1775 was "Americans must not have guns because they are untrustworthy"

When one's defense is weakened one becomes an easy victim.

The Elite continuously compare and associate guns with bad images that are not even closely related to guns. The same with alcohol and driving. If there was a beer bottle cap in the trunk of a car in an accident, the accident is associated with alcohol.

If we supported the Constitution it would devour crime. Unconstitutional gun laws (and most other laws) have been fed to us for so long we've began to like the taste, and now we think they are "constitutional."

Freedom and justice fail when the Rights of criminals are advanced at the expense of the victims.

If the people do not have arms then there is no militia for the security of a Free State. A free People are the Militia. We cannot have a militia if people have no guns. Gun control worked well in Japan, and Nazi Germany, and is still effective in Cuba, China, Iran, Syria, etc. The defense of freedom depends on Americans owning arms. If all the small arms were removed from the American people the armed defense of the U.S. would be cut by over one half. The people own more small arms than the entire U.S. military forces combined. This is a respected position by our foreign enemies, and a designated target for our enemies within to eliminate.

Human Rights

Tyrannical governments barbarously oppress their people. No government has the right to abuse power and control over their people. Inhumane governments criticize other governments, isolate their people, and immunize themselves against convictions and intervention.

Criminal acts by all governments must be judged by their infringement on peoples' rights.

Religion has oppressed, tortured, destroyed, and murdered more people than all other causes together. Even now, religion can still oppress freedom. When morals are above rights, rights take a second back seat. When our freedom is completely devoured, we will have no right to have morals. Morals took a first seat in Nazi Germany. Hitler's morals sat on many rights. Morals are now taking a first seat in many countries, countries like Iran, Libya, Cuba, and even the United States. Laws made to effect morals are really designed to mitigate our rights, infringing on the Constitution.

All the "good" morals in the world aren't worth one millionth of the first ten Amendments. Morals are like opinions, each one is different. We all have different morals, that's a given, but we all must have the same rights, and they are taken.

Laws directed at morality are not only unconstitutional, but also immoral.

Our politicians and legislators are placing emotion above freedom. In our courts the lawyers are placing emotion above fact. In our law enforcement, police are placing emotion before law. In our churches and synagogues our leaders are placing emotion before faith. Even in our media ads, emotions have replaced truth.

When someone does not respect another's emotions they can still retain respect for them. When someone does not respect another's rights they bear malice, and are hostile toward them. Morality laws are a fashion of the Elite, an excuse to force one group to comply with another group's whims.

Loyalty

If we are loyal to our God, then all other loyalty falls into place, including religion, family, friends, country, state, town, school, etc. Loyalty to government should be measured by the government's loyalty to its people. Good government (if that's possible) has loyal people, bad government has no people. Whatever we choose must be positioned within our conscious and pledge.

Media

The media, of course, is the main deception control for brainwashing the American people. They deliver twisted truths, straightened lies, and contrived facts. The FBI is also very good at manufacturing false evidence. The Elite own the media. News papers, radio, etc. not owned by the Elite do not last too long. They are either bought out or drummed out. The media insists on the rights of Amendment I for themselves, but pressures and propagandizes for the mitigation of all other rights, especially Amendment II. These selfish acts help unbalance the Constitution.

It seems American people do not like the truth, for the truth hurts, and Americans do not like to be hurt, physically, emotionally, and especially egotistically.

The media has been an effective tool of deception for oppressing the Rights of Americans. We are gullible, lazy, and apathetic. We truly think that what we don't know won't hurt us.

Ignorance devours freedom.

Since the first newsman spoke the media has deceived the people. The media continues to encompass us with propaganda and lies. We seem to love lies. We love to be brainwashed, even when we think it's impossible. We spend many hours a day being brainwashed, expressing trained opinions, and buying lousy products advertised on TV.

The media takes a human interest story, investigates it and appears to solve the problem for the person. This may be great for the person, but very bad for those who have been sacrificed for publicity, and advertisement fees. This type of news is quite deceiving, it tells little of the true facts, and the news effects that are overwhelming to the innocent majority. If the media were the prosecutors and the public the jury, our prisons would be over packed with the innocent. The innocent would be prejudged, jailed, and the criminals would be free on the streets. Just like it is now.

The media recounts but little truth, since the public fancies large lies.

The media does reveal some truth, just enough to lead us on to believe the rest of the lies. We seem to believe a large lie if it's attached to a little truth. We also reject a lot of truth, if it's attached to a small lie. Therefore, we fail to realize another side of the story, or fable, which is quite often omitted. When the media is accused of a wrong doing they cry freedom of press, and turn viciously on their accuser and adversary.

The media sometimes stumbles on the truth, but it will normally get back up and continue on, fearing it may trip again.

Sympathy, sadness, disheartening, heartbreaking, hardships, and wrong doings are some of the reflections the press misuses to blind and bamboozle Americans. Of all the news in this world, the media reports about 99.99% bad and .001% good.

Nuclear Power

Since 1972 the people of the United States have saved over 500 billion dollars in oil by using nuclear power, and is still saving electrical customers about 26 to 30 billion dollars a year in oil costs. We save about as much in costs associated with pollution due to not burning fossil fuel in power plants. One of the more desirable features is that this money is not paid to the Arabs. The money saved by not buying fuel oil should be used to build additional nuclear power plants, subsequently spending this money on U.S. goods and salaries. This is one reason that anti-nuclear factions are given millions of dollars to provide barriers to nuclear power. There are other factions, like the Arabs, Russia, oil companies, etc. which are spending billions of dollars trying to deplete American oil, to insure the U.S. is dependant on foreign oil.

The U.S. should buy and stock more oil for extra reserves. Then when times of trouble comes between the U.S. and any other nation we would be in fat city with oil, preventing squeezing by any enemy, or "friend." But, our leaders are too involved with the Elite, who control everything, including oil and its prices.

Supreme Idiot Criminals

They are judges, but they can't decide why.

The Supreme Court idiots (SCI) did not sustain a reverse discrimination case, and they brushed off an important drug testing case. I sure hope the SCI don't have to decide on gun control. If they do we may as well pack our bags and go to China. Drug testing is definitely against the fifth Amendment. This is a very dangerous fashion. It is abused by companies, employers, and government agencies.

Why don't we have drug testing for our "elected" officials? A drug test forces one to waiver the Constitution fifth Amendment and testify against oneself. Drug testing is a very good example of giving up freedom for safety and security and then losing all three. Drug testing is out of control, it's spreading faster than AIDS or the flu. The true farce is it started by using public safety as an excuse and now has spread to non critical job personnel. How critical is a baseball or football player to public safety? Some of the reason for drug testing is money. The main reason is control. Our SCI support this unconstitutional sham, and the American people follow like little lambs. The SCI has not based its decisions on the Constitution in two hundred years. Their decisions are for the advantage of the government (WO).

Americans are obliged to defend our Constitution, because the SCI won't. The SCI does not give favorable decisions for the first ten Amendments. It appears they are controlled by the Elite. Therefore they also are quislings.

There is nothing wrong with freedom, it has built a great nation. This great nation is being degraded to an enslaved nation, by design. As freedom goes so does safety and security.

The burden of proving that a law is unconstitutional should not fall on the private Citizen. A law is conceived, written, put through legislative procedure, passed and enforced. An individual must fund the challenge to the constitutionality of this law. We fund our lawmakers to make laws, then we fund our courts to rule on them. Something is very fishy here.

A law should be made, submitted to the SCI for review, accepted or rejected as constitutional, if accepted then passed and enforced. Of course our SCI will accept any law the Elite require.

It seems our SCI have gotten priorities bass ackwards. Nov. 13, 1989, they stated that drug and breath tests, (which invade one's privacy and body) does not violate invasion of privacy, and is not testifying against oneself. Earlier they stated that stomping and burning our flag is freedom of speech. The SCI's credibility is zero. There is no doubt they are the quislings of the Elite. The decision of flag burning shifted the reaction that would have happened if we realized they also allowed private property and funds to be confiscated, without warrant and oath, but only on suspicion, to prevent one from financing a defense council.

The SCI are quislings, and traitors to the people (Citizens) of the United States of America the twelfth. It is a very sorry, sad thing that the SCI has failed American Citizens.

It takes much less money and effort to make a law than to unmake it.

We are now receiving the returns from the apathetic attitude manifested from many years of media brainwashing. Our governmental control is moving steadily towards the Chinese type of control. If you doubt this, try to understand why the SCI allows the government to confiscate our property and assets to prevent us from financing a defense, against this same government. This action is in direct violation of Amendment IV. The media reacted "angrily" to the SCI's decision to allow burning of the flag, yet there was not one whimper put forth scorning the decision to allow seizure of property. This we expected, because the media are quislings of the Elite.

The Elite "Bilderbergers" decided a few years ago, at one of their secret assemblies, that they would increase control of the people through environmental controls. Why would we be surprised that global warming raised its ugly, snake like, dragon head. This means they will convince us that more laws are needed, or we will all perish. They are attacking our reluctance with every entertainer and sportsmen they can buy. We know the SCI will backup any unconstitutional laws the Elite create. The SCI is another dangerous enemy of Americans.

Political Notes

I,   , as a nonresident alien, offer as barter the benefits of my labor and knowledge, receiving all remunerations and compensations as wages without prejudice.

All crimes, if so committed by the accused, are a direct result of duress and coercion from the adverse interest of the State.

An alleged breach of agreement by the accused, whether conscionable or unconscionable, is a prejudicial and reversible error of the State and its prejudice for the assumed parts of agreement. The State assumes, by no fact or evidence, the accused is subject to entrusting under its jurisdiction, and the Uniform Commercial Code.

Illusion of benefits. How can the government give benefits? Is the government of the people or is it self sovereign? As our government is incorporated it is a fictional person. Can a fiction give benefits? Can a benefit be a liability? Can a liability be a benefit? Our government is a fiction and a liability. It for certain our government is not a benefit. The government cannot legally give a benefit to any person without the consent of the people.

As the government claims it is giving benefits to the people, it is admitting an illegal and defacto performance. When a benefit includes a title of nobility, the government is also violating the Law of the Land, Constitution, Article I, section 9, clause 8, and section 10, clause 1. These are illegal actions whether they occur in Florida or Washington, D.C.

A preamble national, common, sovereign person has the right to the sovereignty of the United States of America the twelfth whether he "resides" as a "citizen" in any state, territory or the district of Columbia.

The turncoat Supreme Court (SCI) has allowed the burden of proof to fall on the innocent (accused) rather than the courts.

In our present colorable courts, the accused is not a defendant, but a victim.

Responsibility

Medical facilities and doctors charge high prices because of insurance. In turn, insurance companies over charge us. These same insurance companies also create laws for traffic, industries, medicine, alcohol, tobacco, etc. These laws are unconstitutional and create many hardships on Americans. If all insurance was outlawed, medical cost would be reduced drastically, and self responsibility would return to Americans. How many fathers would let their kids drive wildly if they had to pay all the damages caused by an accident? And if we knew that we must pay for repairs and medical cost (which would be a little lower) as remedy, as assessed by a common law court (which is what we are supposed to have in a free country), or be in contempt of court, we would be very cautious, especially when driving.

In true common law, insurance companies and their greedy lawyers would disappear. People would drive safe, and medical and auto repair costs would go down. Our colorable traffic laws would not be necessary, eliminating crooked judges and greedy lawyers.

Money institutions would be reluctant to lend money without insurance. We would need to save our money to purchase a car. Automobile prices would go down, banks would have much less control over Americans. More important, when we paid for a car we would receive a true title, not a certificate of title.

The elimination of all insurance would begin many, many positive events in America. The illusion of a deficit would disappear. But, the most desirable effect would be a prodigious reduction in the grip of the Beast on our freedom.

The banks offer us credit so we may purchase real property or goods, then when the credit is due and not "paid" the banks confiscate the property or goods, instead of the credit. The banks are stealing our property, our rights, our freedom, our senses, and our lives, in exchange for a promise (false) of material happiness of ownership.

The banks are of satan. Thus, we give our souls to satan for unfulfilled promises. Satan steals our soul through lies and deceit.

The statement on your paycheck stubs "Taxable Income" places your employer in conspiracy with the IRS, which is an illegal international corporate agent. Therefore, the IRS and the employer are in violation of the United States Code

Title 42, Section 1983, 1985, 1986 and United States Code Title 18, Section 241, 242.

Employers also force employees, at their own expense, to comply with ANSI rules which are not legal regulations or codes. Therefore, the employer is again in violation of United States Codes.

Some employers force their employees to qualify and become Firemen. This is forced on employees against their will, not in any contract, and with no compensation whatsoever. Fire fighting is one of the top most dangerous work environments.

Checkered

Irish-American, Italian-American, African-American, what country is Africa? Is everyone so ashamed to be called Americans? I am. The rest of the world's people think we are arrogant, rich, uppity, and stupid. They think we're stupid because we let our politicians ruin our freedom. Yes, they are right, we are so wrong.

Why are some people called black and others called white, and even others called yellow? There is no such animal as a white person, or a black person. We are all brown, different shades, but still brown. The Elite and their quislings want a distinct line drawn between people, and what better way than to divide us by colors, opposite colors. We are (believe it or not) enemies of the Elite. They want us divided. This way we'll bicker among ourselves, rather than turn against them.

The Elite not only call us black and white, they also, secretly, call us slaves. Although, we almost have a decent standard of living, we are still considered the richest slaves in the world. The Elite are designing to reduce our standard of living. They want all the people in the world to have the same living standard. They will not bring the rest of the world up to ours, so we have to go down to theirs. Think about how they are doing this. Jobs overseas, layoffs here, higher prices for everything, cost of living goes up, buying power goes down. Also the laws that they have forced on us, and the laws that they will force on us through environmental controls.

The conspiracy to further enslave Americans is so wide, so planned, so involved, by so many, it is almost impossible to recognize it. Especially if you're brainwashed into believing the politicians. The more that people do not believe there is a conspiracy, the larger this conspiracy grows. This is called a conspiracy spiral. Think about everything that is going on about us, to us. Our first ten Amendments mean nothing to our government. They are discarded in the name of wars, drug wars, terror wars, etc., and now we will have a "war" on environmental effects for global warming. Imagine how many more "Rights" we are now going to lose, along with our limited, but highly taxed wages.

Oh, yes. As long as we think of ourselves as "black and white" the Elite are happy. Our divided colors need to fade into brown, dark dark brown, dark brown, medium brown, light brown, light light brown, and yellowish brown. That is what we really are. We are not like the black and white checkered auto racing finish flag.

After thoughts (Who Thunk It?)

There's not a penny's worth of difference between the Democrats and Republicans. The democrats degrade our morals, security, and freedom. The republicans degrade our standard of living, security, and freedom. One side attacks us from the left, and the other from the right. Law enforcement attacks us from the front, and the justice department attacks us from the rear.

We are surrounded by evil, corrupt, vicious, criminal traitors, assaulting us with deception, hypocrisy, betrayal, treachery, shams, and hoaxes, delivered through a gigantic covert conspiracy. This super secret conspiracy is an assembly of world leaders who meet periodically to devise and construct methodical means and reasons to control and tax (taxation is control) people of all nations, especially Americans. Americans are the easiest to control because they believe they cannot be controlled or brainwashed. Oh man, are we both.

Storms and hurricanes are not the real cause of billions of dollars in damage, the people that build and maintain the cities cause this damage. Hurricane Katrina would not have caused the amount of damage if it hit Miami, rather than New Orleans. New Orleans has always been defenseless, and susceptible to storms due to its layout and stupidity of design. Who builds a village in an ocean overflow drain? Dumb asses like the ones who built New Orleans. Miami is very susceptible to storm damage, but it is not below sea level.

Guess who pays the 125 billion dollars in damage. You know it, we do, the working people, the retired people, and the poor people, but not the rich people or the government, or the oil companies, or the insurance companies, or the banks. Not even state or local governments pay out funds. It all comes from us, the brainwashed sheep. When the companies and the government lose just 1% of their gigantic profits, they whine and raise prices or taxes.

We "give" a very large percent of our allowance (salary) to banks, insurance companies, oil companies, and the government. We subsidize (pay taxes) the government and all the large companies, to keep them rich, while we get poorer and poorer.

If we worked 40 hours a week earning 1000 dollars, we would have only about 10 dollars at most for "spending" money. Adding up all our taxes (income, excise, sales, property, and all the hidden taxes like licenses, fuel, insurance, and food, we pay them well over 90% of our salary. And now they have found a way to tax us on the air we breath, with "global warming." About 70% of our salary goes to straight up taxes, and the rest goes to interest on mortgages, loans, charge cards, etc. All the above fees and payments are all taxes disguised as benefits.

When we are given a "benefit" we must submit us and ours, to the federal government, which is a corporation. This makes us corporate personnel subject to their rules, constitutional or not.

We have no United States of America the Twelfth real courts in this country. We are still under Roman law. The court systems we have are equity and Admiralty. We have no true common law courts. Lawyers are not allowed in common law courts. Any court with lawyers is not a true Constitutional court.

The major cause of the loss of personal responsibility in this country is insurance. Insurance removes caring from drivers. If all insurances were eliminated responsibility would return. Banks would have the greater risk, not the people. Government gets less kickback and bribes. Lawyers begin to disappear. Prices go down (especially medical and drugs). Peoples' freedom begins to return (until environmental laws and taxes are invoked).

The greatest scams:

> Federal government and its agencies (including elections).
> State government
> County government.
> Local government.
> Banks.
> Insurance.
> Oil.
> Laws under color
> Lawyers.
> Entertainment, news, TV, radio, sports.
> Education.

This list goes on and on, but it all hangs on the federal government and all its scams and conspiracies.

Lies, the world is all lies, and so hideth satan.

Disasters

As bad as it was, "9-11" was a drop in the bucket compared to other man made disasters. The attack on the world trade center knocked down a few buildings and killed about 3,000 people. The A bombs on Japan killed over 100,000 people. The Russians (and Germans) killed millions of Polish people. There are many other disasters such as wars, fake wars, police actions, etc. that killed millions of people, and imprisoned many, many more. About 90% of all prisoners in the US are political prisoners.

Man made disasters cannot begin to compare to disasters created by nature. Below is a list of some of the natural causes of disasters:

avalanche	hail	rogue wave
blizzard	heat	Storm/hurricane
cold	landslide	tsunami
disease	lightening	sunspots
draught	locus/plague	tornado
earthquake	meteorite/asteroid	volcano
fire	mudslide	waterspout
flood	rain	wind

I'm sure we can think of others, but this is a start.

The "Elite Bilderburgers" are the One World Order. They are satan's
partners, associates, accomplices, fellowships, alliance, cabal, junta, and especially the federation (including our federal government).

When our politicians aren't feeding us straight lies they're nurturing us with twisted truths, the same as our scientists.

Cosmology and Cosmetology have much in common. They both cover up the truth with a multitude of lies.

We believe we are so smart, and know so much, yet we are so stupid and know absolutely nothing, especially the truth.

How can we know the truth in a world of lies?

If we added up everything we have and what we are worth, multiply it by a million, and then divide it by infinity, we would have exactly what we are worth … nothing, zero, zip, naught, aught, nil, goose egg, whatever, nihility, nonexistence, void, empty, vacuum, vacant, and only dust (at most).

If there is something less than nothing, we are it. There is one thing we are and that is egotistical giants, and proud of it. If you don't think Americans are stupid, just watch the way we drive. The most considerate, lovely, intelligent person will get behind the wheel and transform into a raving idiot. What a magic trick we perform.

Americans are no longer real Americans. True Americans fought for their freedom. Now, we cowardly, false Americans ask to be enslaved rather than be inconvenienced. We say "Give me money and let me be your slave." We say "Give me security and let me be your slave." We say "Give me safety and let me be your slave." We are all "colored." Color has absolutely nothing to do with race.

We are so stupid we now say "Take my liberty and take my rights, but please lend me money." We give up our real rights to borrow false (fiat) money. Our weaknesses are well known by the world order.

We solemnly promise the town, the county, the state, and the federal governments to work at least fifty years to support them and their wallets.

The same entity that controls the Supreme Court controls the ACLU, many Churches, the Senate, the House, the Executive Branch, the States, the Banks, the schools, and hell.

It only costs the Feds 4 cents ($00.04) to make a dollar bill of any denomination. If the total face amount for each set of bills printed is $1,000,000 we will find:

Bill denomination	number of bills	cost to make
$ 1.00	1,000,000	$ 40,000
$ 2.00	500,000	$ 20,000
$ 5.00	200,000	$ 8,000
$ 10.00	100,000	$ 4,000
$ 20.00	50,000	$ 2,000
$ 50.00	20,000	$ 800
$ 100.00	10,000	$ 400
Total	1,880,000 bills	$ 75,200

Seven million dollars costs only seventy five thousand two hundred dollars to print. This cost decreases as more bills are printed, and reused.

If you earn $ 50,000 a year, and pay out over 60% of your salary in taxes (i.e. income, sales, excise, property, licenses, etc.) That is $30,000 given to the "government." With the $20,000 left you pay for required insurance (i.e. fire, flood, auto, life, etc.).

The Feds pay you $0.04 per dollar you spend, or work for. The average dollar bill you carry is the $20 bill, the $10, the $5, and the $1. Add these up we get 4 bills for $36. Four bills cost the Feds $0.16 to make. For an average cost of 16 cents for 36 dollars, they pay you $50,000 a year, but it only cost them $222. They just saved $49,778 on your labor. Then you pay them back $30,000 whole dollars in taxes. So far, the Feds actually make money for your labor (over $79,778). It really gets worse because they use the same dollar bills to pay you again and again, etc., making even more profit on your labor.

It even gets worse when you charge credit and pay interest. Interest and insurance are paid to the same factions that we pay taxes to, making them a covert tax.

We pay more than full price for government's labor, but they pay us very little for our labor. They also pay almost nothing for the benefits they "give" us. Are we dumb, or are we really dumb and very stupid. I know we are political cowards and very lazy. How can we be fooled so easy?

When you are forced to buy insurance it becomes a straight up tax. Then, when you make an insurance claim they cheat you. When they finally pay a claim it is at the 4 cents per dollar rate that they are actually paying; and many insurance companies claim it is paid as interest (making it taxable).

The following is a list of lies used to deceive and control us. When any of these lies are questioned the wrath of the world order will fall on whoever speaks the truth.

Lies to Americans

aircraft contrails (chemtrails)
Amendments XI, XIV, XV, XVI, XVII, XIX (are not Constitutionally ratified).
Americans are not brainwashed
Bible Code
big bang
Bill Gates, Microsoft
budgets (local, state, federal)
carbon dating
"citizen" (Citizen)
civil war
Constitution
crime doesn't pay (ask Rockefeller, Rothschild, FBI, politicians, etc.)
earth creation, history
DaVince code
Einstein (E=mc2)
elliptical orbits
evolution
expanding universe, big bang
"fossil fuel"
freedom of Americans
freedom of information act
fusion powers stars
gas prices
global warming
government agencies (FBI, CIA, FEMA, IRS, etc.)
government serves the people
gravity originates from mass
holocaust

home of the brave
income taxes
interest rates, Federal Reserve
judges, lawyers, prosecutors, etc.
Kennedy assassinations
land of the free
laws under color (federal, state, county, local, traffic)
light interference (patterns)
light speed limit
marriage, "children" (state custody)
money (dollar bills)
moon's gravity
moon landing
Murrah building bombing
national security excuses
no cold fusion
ownership (home, car, property, children, etc.)
ozone layer hole
Pearl Harbor surprise
property (land)
rapture
space warp
time dilation
TV, radio, newspaper advertisements
TV, radio, newspaper news
UFO excuses
United Nations
United States/United States of America
US deficit
voting/elections
Waco holocaust
wars (on drugs, terror, or anything)
Watergate (Nixon)
we are a free country
World Trade Center attackers

The Greatest Liars

Note: This list is far from complete.
advertisers
doctors
entertainers
governments
Kenites
judges
lawyers
media
news personnel
politicians
religious
scientists
sports figures

Politicians have found another way to "screw over" people and get their own wallets filled.

"EMINENT DOMAIN" The "Right" of the sovereign or state to take private property for public use. Our wonderful Supreme Court criminals decided that the state can take private property for private use. The lie is that the property will bring business to the people and enhance the community. The truth is that the property is practically stolen by the state and millions of dollars go into the politicians' bank accounts (mostly from contractors). Most of the victims are poor and cannot afford to fight them in court, or payoff the politicians.

Whoever (or whatever) is piloting UFO's have some very high connections in all countries. Their cover ups are thorough and complete. Whoever, whether they are earthlings, satan's minions, or aliens, that have their designs on the control of Americans and other peoples are very, very intelligent, and have been around for a long, long time, and they can plan ahead for thousands of years. Of those who may rule this earth it is probably satan rather than aliens.

Our government prints fiat (counterfeit) money, forces us to use it, then taxes us for using it. Do the math ... it cost the feds only four (4) cents to print a dollar bill, no matter what denomination. When we protest we are thrown in prison by the Federal kangaroo court.

It certainly appears that our Federal Government is totally and traitorously a false government. Our false government never tells the truth to the people, because there is no truth in them. The huge problem is that we the people do not know the difference in truth or lie. What do we have to compare a lie to? It is worse than that, we the people reject truth and follow the lie.

Federal agents have no Constitutional jurisdiction over Preamble Citizens. We have been deceived and lied to, to pull us into their corporate government. All that we know, and believe about our freedom are lies. It seems that the more they lie to us the more we love it.

Think about what our "government" is doing. They won't let a man or woman marry more than one spouse, but they'll let doctors kill babies. They'll put you in jail for shooting a dog or cat, but encourage women to kill their unborn offspring. The government is all hypocrites, pretenders and liars, when we do not believe their lies they throw us in prison.

If someone counterfeits counterfeit money are they criminals?

Never let your education destroy your intelligence!

The health of Americans (and others) has been slowly attacked through chemistry and electronics. Everyone knows that radar (electromagnetic energy (EME)) can cook or destroy flesh. If a concentrated amount can kill, then a smaller amount can injure. Over time EME, even in small amounts, can injure or kill flesh. The earth's surface is jam packed with human created EME. This is far worse than "global warming."

Thus, we are slowly being destroyed through our injuries caused by EME, and *we* are paying for it. Many diseases are caused by EME. It only takes a small amount to upset the chemical balance in our body, especially our brains and endocrine systems.

We may be created equal, or even born "free," but almost all of us will die in servitude. Slavery was not abolished, only renamed "citizen" with a small "c."

There are several ways to control the people of a country, or even the world. The most effective is economics. Another is creating false laws. Another is taxes and licensing. Another is "brainwashing" through the media and education. The best way to control us is to use all means available.

Reducing our power to buy by cash forces us to buy through financing. And, we know who owns the banks, the media, and the insurance companies. The World Order (also the Global Business Plan) has been using all these controls for over 200 years; although, it is much easier now, due to modern media, which deceives millions and millions at once. Control is gained through small steps at a time, raising little suspicion. The suspicious ones are labeled "nuts" or criminals, and usually thrown in prison.

Our "government" does a very poor job of hiding the fact that they're hiding facts. Although, it is quite evident that the government is very good at hiding true evidence, and creating its own evidence, especially the FBI.

We (Americans) pay a much higher percent tax from our "income" now than Americans did in 1775, and they fought a war over it. Americans used to be knowledgeable and brave. We (Americans) are the dumbest, stupidest, laziest, most cowardly people on earth. This is because we are the most gullible, foolish, greedy, and brainwashed.

By the way, the Global Business Plan stated at a "Bilderberg" meeting a few years ago that they will control the people through the environment. "Global Business through Global Warming." We will be coping with many more "laws" and taxes.

We make our living by what we get. We make our life by what we give.

Energy is much more than you think. It's more than light, or heat, or power, or force, etc. Energy is life itself, and even more than that, it is really "after life," our spirits and souls. As our bodies "die" they become dust, but our spirits (energy) are set "free."

Pure energy, attached not to mass, can travel at infinite velocity, covering infinite distance, instantly. Therefore, the vast distances of the universe, and creation are only a "step" away. Spirits and souls can be everywhere at the same time, with no interference from each other. There is no time in Heaven or hell. Time is manmade.

When our heroes earn millions of dollars through lying and depicting lies (acting), and believes winning is more important than someone's life, we are super brainwashed, very stupid, or of satan's minions.

The posting of the Ten Commandments at a government building is a waste of good granite, because inside lurks judges, lawyers, and politicians.

Hypocritical

We have laws that imprison people for killing a dog or cat, but we allow people to kill their unborn people.

Our "elected" politicians call us free, yet they make "laws" to enslave us.

Lawyers make laws to separate childhood from adulthood, yet they put youngsters on trial as adults.

Kids want adults to be tolerant of their ways, yet they toss aside any consideration of past generations. Kids reject the meaning of past ways. Soon they will enjoy rejection, because the earth will not end before they have their own kids.

There is not a nickel's worth of difference between kids now and kids 2000 years ago.

There is not a penny's worth of difference between democrats and republicans. One's heads and the other's tails, connected solidly together, made of the same cheap metal, created and controlled by the Feds.

If you work for the truth, you do what the truth needs. If you work for the money, you do what the money needs.

The Federal, State, and Local governments cannot tax anything they do not own. Think about that for awhile.

If evolution is for real, we should have found some skeletons showing the change of an animal from a few thousand years ago to now. Doesn't it seem a little weird (and suspicious) that we have never found a true missing link—from any species?

Did you know that yoga is a pagan ritual? Many of the yoga positions imitate false gods.

The extreme left are schizophrenic, the extreme right are paranoid.

If you are paranoid about the government, you are probably more knowledgeable than most Americans, and less brainwashed. When you are paranoid about something that is true, you aren't paranoid, but wise.

It is much smarter to be paranoid of the heat than to fall into the fire.

Why do we believe news about the government? Our government has not told us the truth for over 200 years. Why do we believe politicians? They have never, in history, told the truth. Since the first politician uttered ugh, ugh (me help you) there has been no truth told.

If you would look into any popular American English dictionary it would describe "politic" as crafty, unscrupulous, cunning (similar to "suave"), etc. It would also describe "politician" as one who is interested in personal or partisan gain and other selfish interests. "Suave" is controlled, refined manner, with artful management, tact, and shrewdness in gaining an end. Yes-sir-ree, we sure do have suave politicians in our governments, and our businesses.

If American judges are not truly elected then they are not true judges. Appointed judges, such as traffic judges, magistrates, federal judges, etc. are not real judges, they are a lie of law. These "judges" are true criminals, worse than politicians.

We are super-brainwashed when our heroes earn millions of dollars through lying and acting. It seems like there are about five billion people that are super-brainwashed. The other one billion people are still too young.

The ATF, FBI, IRS, etc. are strong-arms of the Global Business Plan leaders, the WO masters.

Governmental conspiracies have existed since mankind first partnered with each other many, many years ago, even before we had organized language. And, has continued ever since, including our own present (and past) government and large businesses. I would not be surprised if aliens controlled the governments. There are many clues, and much evidence, for this suspicion. Think about the UFO cover ups. Anyone who runs the world governments will go to extreme lengths to protect their identity. Also, the countries that tend to resist and oppose the global business system (new world order) are eventually "repatriated," one way or the other (Panama, Iraq, Afghanistan, etc.).

If one looks at the history of nations and societies one will find that these societies decay as the gaps between the rich and the poor widen. This is not just a modern phenomenon, it has happened in many ancient societies such as the Roman, Aztec, Mayan, etc. This gap is now happening in the US, Canada, England, and most of Europe. The rich/poor gap is prominent in third world countries such as Mexico, South Africa, Peru, and uncounted others. The global controllers seem to want to lower the standard of living for the working class in the US.

Doesn't it seem like the more educated people are the more stubborn, egotistical, and stupid they become. The higher their degree the dumber their temperature. Is this due to the brainwashing that is imbedded in education, or is it that the more the brain absorbs the greater the stupidity, or both? There are many, many, many lies taught in our education system (the list of lies to Americans names a few).

What we think are good ideas, good laws, public policies, etc. are actually not so good. What we think is bad is actually not so bad. Our sense of truth has been turned around to brainwash and manipulate us, by the world order (WO).

The "New World Order" itself is a lie. It is not really new, it is thousands of years old.

We have "laws" for a "reason." Unfortunately, the reason for most of these laws are not for the good of the people. As we now realize laws are made for lawyers, not real people; therefore, about ninety nine percent (99%) of our "laws" are fictitious.

Below is a list of false Laws.

act	law
bill	limit
bylaws	mandate
canon	manifesto
code	order
constitution	ordinance
custom	policy
declaration	proclamation
decree	public policy
dictate	referendum
directive	regulation
doctrine	resolution

edict	rule
guideline	standard
injunction	statute
instruction	under color

These and other "laws" have been taught to us, and forced on us as real laws. The only one that can make a real Law is God. We have been deceived. We are being deceived. And, we seem to like it, we keep asking for more and more "laws." There are about four to five million "laws" in the United States, and the number grows with leaps and customs. The WO now wants Americans to comply with international and foreign laws. They are forcing American service personnel to serve under the UN flag. This is truly not Constitutional.

Of course everything the WO does is unconstitutional, and Americans love it. Therefore, we have no Constitution.

Entertainers, sports figures, politicians, and billionaires do not make millions and billions of dollars because they were smart, but because Americans are stupid. Remember, stupid is terminal, and so are our "rights." Because of stupid, our Constitution is no longer valid, and effectively terminated.

Americans, more than any other people, are fed and overstuffed on bull crap and brainwash info, continuously. Every time we listen to or view the news, sports, ads, and politicians we absorb lies. The news programs support the police state. If you think we are not in a police state just try to insert your rights into the conversation with a cop. The police are the authority (the boss), not the Constitution. What we view on the TV programs are nowhere near the truth. *Only if all ads were all true, what a wonderful world we would have.*

Those who are guilty of many things should not accuse others of anything.

Remember, once again: when you give up your rights and freedom for security and safety you lose all of them.

The greater the conspiracy the harder to prove it.

Should I pay $8.00 to sit in a smelly, dark, germ infested seat to watch a large bright, noisy screen, and pay exorbitant prices for lousy food and drinks just to see and listen to brainwashing Hollywood lies for two tortured hours? The only real entertainment in a movie theatre begins when people start shushing each other.

Did you know that the US Government makes it very hard to adopt children, just to encourage abortion?

Wakes and funerals are for the family, but the mass is for the deceased.

Beware, for those who want abortion; they themselves will eventually be aborted.

Movies, like Harry Potter is another satanist doorway to the occult. Occults create lies, murder, marriage abuse and divorce, abortion, birth control, suicide, gayism, and un-numbered other transgressions.

Those who insert limits into physics (i.e. light velocity) do not believe in God, nor do they truly know physics.

Again, if evolution was true then there should be transitional evidence, but there is not one valid piece of "missing link" of any species. If evolution is true then there should be thousands and thousands (probably millions) of human and animal transitional evidence.

Why is evolution going to be taught in public schools? There are factions trying to eliminate home schooling. This would afford more lies to be taught in public school.

Our police state is becoming more and more, day by day, just like WWII Germany. They throw us in jail because we don't like someone (called a "hate crime"). They throw us in jail because we won't tell them our name. They throw us in jail because we won't tell them what we might know. They throw us in jail if we don't have our papers. They throw us in jail if we don't tell on others.

For some reason the US government seems to design means to shorten our life spans. They inject us with vaccines containing ethyleneglycol, phenol, formaldehyde, aluminum, thimerosal, neomycin and streptomycin, mercury, and claim it prevents diseases. They inject our drinking water with poisons and toxins (fluorides), and claim in prevents cavities. They inject chemicals and toxins in our atmosphere (factories and chemtrails), and claim denial. They inject laws into our freedom that prevent us from attaining good, better, and cheaper medicine, and claim safety. They bombard us with thousands of lies through the media, and claim it's entertainment or news.

If one searches the Presidents Executive Orders one would find that the American people are considered the enemy of the US government. The US government considers the American people subjects and guinea pigs. Remember the United States government is not the United States of America, the twelfth government. The United States is a corporation and we are it's subjects. Our names are all in capitalized letters, depicting us as small "c" citizens.

The United States of America, the twelfth is the Constitutional government, or should we say **was** the Constitutional government.

The tactics and conversations being used to convince us of global warming would not be necessary if there were true scientific grounds for global warming. Soon they will claim a global disaster and emergency measures to be taken. Then there will come the laws and taxes to increase the control the World Order ("Bilderbergs") has over us.

"Close" counts in horseshoes, hand grenades, hurricanes, halitosis, and federal law.

China vs. Mexico

If our government really wanted to stop Mexican immigrants from entering the US it would have been done years ago. One way to end this illegal influx is to take away some of the jobs outsourced to the slave driving nation of China and assign them to Mexico. This would have some major advantages, such as:

1. Cheaper shipping
2. Reduce child slave labor
3. Reduce funds to communists.
4. Reduce illegal immigrants.
5. Reduce funds to terrorists.
6. Increase American/Mexican relationship
7. Reduce requirements for Border Patrol
8. Reduce chance of terrorist infiltration
9. Increase Mexican standard of living.

Alas, our beloved, wonderful, honest, trustworthy government has different plans, to continue our enslavement and lower our standard of living. Once again, the greater the conspiracy, the less conspicuous.

If you don't believe that all politicians are criminals just listen to the political advertisements before elections, they all snitch on each other.

Judges don't judge, they condemn. Judges who are not elected are truly unconstitutional, therefore their decisions are not constitutional. They do not have the authority, but they have the backing of the unconstitutional law enforcement. Federal judges are definitely not constitutional.

When federal law overrides state law, it is proof something is not correct with our constitutional system. As a matter of fact, we are not under the constitution, but are ruled by contract law created over 2000 years ago, and strengthened 800 years ago.

The only constitutional law enforcement departments are the sheriffs, and they enforce unconstitutional laws. They have absolutely no authority to enforce federal "laws."

You can not see light going away from you.

Scientists will never understand the true physics of the universe until they realize that gravity does not originate from celestial objects, and that orbits are not elliptical, but sinusoidal. Scientists, in their ego, do not follow their own rules.

Lies we have received from scientists:

> photons have no mass
> orbits are elliptical
> velocity of light is maximum velocity

CO2 causes global warming
expanding universe
space warp
carbon dating
distance to galaxies
no cold fusion

The insurance companies are as greedy as the oil companies, the building material companies, the banks, the computer industries, the media and entertainment companies, the internal revenue (dis) service, and politicians. This is because they are all run by the same people—the World Order.

We truly get screwed from the insurance companies. They claim to lose money from hurricanes, then raise rates (some over 1000%). They do not lose one cent from paying home repairs. They receive great profits from the materials used to repair damaged homes.

What we have been taught about history is not quite true. There are a couple of sample truths thrown in here and there, but the majority is twisted truths, curved fabrications, and straight lies. The meaning of words (in all languages) have changed, and are still changing. Even American history has been distorted.

History is created through lies.

Distortion of history is done for specific reasons, to confuse the "good guys" with the "bad guys." Many of the people we have been taught to be wary of are the original "good guys." Many of the people we have been taught to trust are the original "bad guys." Many evil people still exist, such as thieves, murderers, and the "WO." Many people in prison have been "framed." The WO has created an American Police State, and imprisoned many who have resisted them. The WO has created "laws" that cannot be obeyed, yet they imprison the ones who "violate" these laws.

The WO enforces its rule through fear. Their agents such as the FBI, IRS, BATF, CIA, etc. enforce false laws, unconstitutionally confiscate property, and assimilate Citizens into their corporate laws (making them citizens). There are no constitutional laws that requires American Citizens to pay tribute to, and to render income taxes to the United States corporation. All those thrown in jail for tax evasion, etc. are not criminals, only victims. The ones who send them to jail are the true criminals, and traitors to the American People. We have been hoodwinked, deceived, conspired against, finessed, and tricked into becoming United States corporate citizens. We have not accepted this willingly, knowingly, intentionally, or voluntarily, but stupidly.

Our civil war was not for the heroics, to be valiant, majestic, or even sympathetic, in order to free the slaves; rather, it was to enslave all Citizens, making them citizens of the United States corporation. The southern leaders saw this enslavement in its primary works. They could not fight this in congress, so they withdrew and retreated because the United States of America the twelfth (a democratic republic) was becoming the United States, the corporation. The Civil war was not a war at all, or even close to a war, although bloody as it were, it was only a police action of the northern states to return the fleeing southern Congress to the capital, or the US corporation would be dissolved. Evidently, the corporation won, illegally assimilating Citizens, making us citizens.

When your name is printed in all capital letters it indicates that you are a subject corporation, whether or not you care, having no choice. We are now corporate slaves. Presidential executive orders consider citizens as enemies, and may treat citizens as guinea pigs. The police state we are in was created by the origination of the US corporation, and the "WO."

Americans are more controlled than the Chinese because we believe we cannot be controlled. It is very easy to brainwash people who are lazy, greedy, gullible, know-it-all, and believe that they cannot be brainwashed.

We do not own anything. We pay rent on everything we "own," it's called taxes. Insurances such as health, auto, home, etc. are also a form of taxes. If you think this is not true don't pay tax or insurance on something you think you own and see who really owns it. Remember, the government cannot tax anything that is not theirs.

The property titles you have are not truly titles. When you pay off a car you receive a "Certificate of Title." When you own a house you have a "Warranty Title." The state holds the real vehicle and property titles. We are also citizens (small c) of the state. This means we are corporate subjects of the state. The states are corporate subjects of the federal government.

The federal government is a corporate subject of the World Order (WO). The WO is the subject of the Kenites (descendants of Cain), the Kenites are the subjects of either satan or aliens, or both. We are locked in slavery, although we do not realize it because they have kept us brainwashed, lazy, and greedy.

Even our earnings are subjects of the state. The federal reserve prints their own money, forces us to use it, then "taxes" us for using it. We cannot avoid eventually using the bank. Chickens, pigs and cows are no longer goods of trade for each other, we are forced to traffic in federal money. A point of information: If a judge ever asks you "do you drive in traffic" say no. What he is really asking is, do you deal in federal goods, which subjects you to corporate laws.

"An Inconvenient Truth" is really a convenient lie, and another huge conspiracy for the World Order. The WO is pushing this global warming lie stronger than they pushed the holocaust lie. They are truly in a rush to control their subjects. They plan to increase taxes, laws, rules, and prices, to increase control of Americans (and others). This gives them even more police power. We are already in a strong police state.

We have laws to protect eagles, dolphins, turtles, even alligators, etc. In the same books, we also have laws to allow killing of unborn people. If a women has an abortion it is paid for by the state, but if she leaves a baby on a doorstep she goes to jail. Abortion is intrinsically evil!

Laws are made for the elite, lawyers, and the government, not for the people. The ACLU and the ADL are just tools for control and another compliance for the WO controlled media to root for, and to continue to brainwash Americans.

Why should we save the earth? The United States is the "greatest" country in the world, and it sucks big time. *Freedom is like food. If one has a lot of money one has a lot of food. If one has no money one starves.*

What we need to save are our souls, not earth. The earth belongs to satan.

Can you imagine how great this country would be if our Constitution was involved. There would be no police state and no Kenite could involve us in any conflict, and all advertisements would be true. The Waco holocaust (a true holocaust) could not have happened.

We can not have justice until we change our justice system.

It is absolutely useless to use science to confirm or deny religious beliefs or history. It proves nothing but lies, either way.

If one thinks that Americans aren't stupid just look at the salary differences they support. CEO's, actors, entertainers, and athletes receive about $100,000 to $100,000,000 or more per year. The working class, tradesmen, and soldiers earn only $10,000 to $60,000 per year. The true shame is that the tradesmen, soldiers, and the working class support these very over paid elites. Stupid? Yes! Yes! Yes!

Greed creates money, money creates power, power creates control, control creates greed. Greed is also a by-product of self seeking traits.

There is a great rainbow above the United States.

Why should we save the earth? This earth is not ours, it belongs to satan and her minions.

Humans have thousands of years experience in politics, science, and religion and we still get it wrong.

We are forbidden to be prejudiced against anyone who serves satan.

When scientists claim they know something for sure, they are either very stupid, very lying, or very both.

Scare tactics are often used very successfully on us "dumb ass" Americans, to help increase control over us.

The United States appears to be the free country, but we are going downhill rapidly. We have corruption in politics from top to bottom. We have a true police state. We are continuously bombarded with propaganda through the media and "education." We are subjects of the golden rule; those who have the gold, rules.

We have about eighty percent (80%) political prisoners in state prisons and about ninety nine percent (99%) political prisoners in federal prisons. This is more than any other nation.

The more we have been taught by the system the more we have been brainwashed. The more we watch television, the more we are brainwashed. The more we are brainwashed, the more we want to be.

The "US Government" never allows the truth to defray an opportunistic theory. It will never let consequences interfere with a lucrative idea. It will never allow the Constitution to defeat a profitable law.

The author(s) of the "da Vinci Code" and other "enchanting stories" are mentally ill, or evil, or both. They are definitely supported by the evil WO.

If

UFO's are real and the governments are hiding the facts, then there are aliens controlling governments.

There are paintings and scriptures depicting UFO's since history began. Aliens may have controlled people for thousands and thousands of years. If we look at history, it appears that events have been manipulated. Important events and inventions have come around slowly and deliberately, not rapidly or instantly. This would mean that those in control have been in control for a very long time.

All our unconstitutional "laws" have been slowly introduced, over long periods of time, little by little, not greatly alarming anyone. But, if we look back at all the "laws" created it is very alarming. By the way, congress does not make laws, they make resolutions. These resolutions are for us corporate citizens (spelled with a small c). Again, when you find your name in all capital letters you are a corporate citizen, subject to the United States Corporation. Corporate citizens are not Constitutional Citizens (they no longer exist). We have all been forced into the United States Corporation, which is incorporated under the "World Order."

There is no longer a United States of America the twelfth. Therefore, there is no longer a Constitution for the United States of America. We have succumbed to the Global World Order. All these rights we demand from our "government" no longer exist. We have never left the rule of the British and Roman Empires.

If you don't think there is a conspiracy to control people, especially American people, then you don't think.

How did we lose our Constitution? How did we get a police state? How did we get a federal government that runs the states? If there is no conspiracy, then it must have just happened by accident.

If you read the "Bill of Rights," then a few of the six million "laws" we have you may begin to suspect something. Or, if you research "under color of law" you may begin to understand something.

Warning: When you begin to understand what has, and is happening to our Constitution and rights, you will become angry. There are millions of angry, but cowardly, people in the US.

The laws of the United States are depicted in Revelation 10: Our "Law" is the angel clothed in a cloud, and a rainbow overhead, holding a little book. This is explained in the first section Book 3.

Phaith & Fysics

All the science and all the mathematics in astrophysics has not increased our knowledge of the universe. Cosmology has done the opposite. It has led us down the wrong street. Science cannot know the dynamics of the universe as long as it continues down the street of cosmic lies. Astrophysicists, with all their training and "education" still do not understand the dynamics of orbits. Scientific egos preclude knowledge. The string theory will probably be accepted because it is too complicated for the general public, although it's close, yet it is still wrong.

From ancient Greece to now, science has been wrong. Each theory has been replaced by another theory, and another, and another. Present theories are also wrong, but scientists are egotistically connected to their training. It can be proven that orbits are not elliptical, that gravity does not originate from mass, and that photons have mass. Scientists will not even listen to anyone that is within their circle, much less anyone that is outside their political "click." The advances made in physics have been accomplished through accident and/or trial and error, not theory.

There is no intelligence on earth. If there is it is outshined by ego. No one can have practical intelligence if they are maintained by their ego. If our intelligence is rated 10, then our ego is rated 10,000.

The universe is gigantic, and God commands every photon. Our universe is full of electromagnetic energy (EME) of all frequencies, color, and strength. All this EME is gravity (see the Universe of Energy).

We are only one very short beat in the eternal song of the universe. Our one note could be skipped and the universe would not notice, the music would still be beautiful.

God created and commands each photon in each and every EME wavelength. Therefore, God is gravity. God is the universe of stability and chaos. The laws of physics are God's laws, not man's, man cannot make a law.

Atheists have attached their agenda to the meaning of evolution. Yet, evolution has never scientifically been proven, it is only a theory. Atheists create false science such as the big bang, time dilation, evolution, etc. to further their agenda. Atheists claim that God should not be in government. They must have their way, because there is no truth in our government.

Government should not be in religion, but religion must be in government. If there is anything the American people need is a little honesty in our government, right now we have none. But, first we need honesty in religion.

Random: anything we cannot predict, calculate, or estimate. Actions beyond our knowledge are "random." In other words, random is a consequence of ignorance.

If we were to guess who rules the World Order, we have a small choice. To choose between the Kenites or the aliens is not much of a choice. The aliens carry the vote because the Kenites probably did not have UFO's. UFO's have been depicted in ancient paintings. If aliens have been around that long, then they have a plan. They would be the only ones to have a long range plan (over thousands of years). But, why do they besiege and imprecate Christians? Maybe Kenites are aliens. This explains many questions. Aliens infiltrate Solomon's reign, and continue to prosper. Of course, Kenites may be aliens.

Who needs laws? Lawyers!

Who thrives on laws? Lawyers!

Who makes laws? Lawyers!

Who gets screwed by laws? Citizens!

This is why the United States is a police state.

The ones who truly need to be tried for treason are judges. Not just some, but all judges. All judges in the US are traitors to American Citizens, including the Supreme Court traitors. The conspiracy to control the American people could not happen without the effort of our traitorous judges (and lawyers). All laws and judges are clouded and have rainbows over them.

Maybe it goes this way: God created the universe. This would include any other life in the universe. The children of satan could come from another world, then took over the earth. Christians do not belong on earth, earth is satan's world. Kenite, alien, satan, all the same because satan came to earth to seduce Eve, and to produce Cain. Adam fathered Able and Seth. Seth was second in line to Jesus.

At one time satan offered the world (Earth) to Jesus.

Ain't it strange that "prominent" people deny the claims of UFO witnesses but still believe the propaganda of our government and politicians. How can anyone believe politicians when they state "we're here to serve you?" There has not been one politician in history that has told more than one percent of the truth. That is quite a record, yet people still trust them to decide and create our fate.

Our government is evil, very evil.

If one would read just a few of the thousands of presidential executive orders one would understand that Americans have no Constitution. We are only subjects of the federal government and its agencies. Our lives (and deaths) are at the whim of our government. We are the cattle that are being led to the political slaughter house.

The World Order (WO), controlling the US government and its agencies, are excellent at three things: taxing us, scaring us, and lying to us.

It seems everyone wants to scare us. Our government, the advertisers, our employers, the religious, and anyone who wishes to influence and control us, all use scare tactics.

Out government lies to us to scare us. They scare us to tax us. They tax us to control us. We are much easier to control and lie to when we are in fear of some sort of a disaster. We are not taxed so that the government can have money. We are taxed so the government can have control. *What they can tax they can take.* The government does not need money, they can print all the money they need. *Money is power, the less money people have, the less power we have.* Our government is evil, very evil.

Americans have:

Eyes to look at mirrors and TV's,
Ears to listen to propaganda,
Noses that cannot smell anything fishy,
Mouths to brag about themselves,
Brains to be washed,
Fingers to play games, or point at others,
Feet to get in the way of their remote controls,
Butts to plop down, and a place to store lead.
Beware those who have the words to up themselves and down others.
How can I judge others, when I'm stupid too?

Those with silver tongues get to taste the gold.
Those with the gold get to taste the rule.

The "New" World Order once stated at a "Bilderberger" meeting about fifteen, or so, years ago, that they would control the people through the environment.

They are now doing exactly that. They have created another great lie and prevarication since the income tax fabrication; "global warming."

They are also creating "climate change" through spraying the atmosphere with chemicals. If you've ever seen the "contrails" that last for hours in the sky, you may assume they are normal. They are actually chemical trails ("chemtrails") that are causing atmospheric change. Real contrails are short and follow aircraft, evaporating behind the plane as it moves.

The WO has been spraying chemtrails for at least 25 years, with very few people noticing. Although, the earth's climate has been changing since the earth began, the WO wanted a little faster change, in their required direction.

The WO will increase their control by increasing laws, prices, and taxes. What they can tax they can control. The WO controls about ninety percent (90%) of our life now. In a few years they will up that from 90% to 95% or more, and people will rejoice because they'll think the environment is safe, and they'll be safe from terrorists.

We will never be safe from terrorists until we eliminate all the federal agencies, and most of the state agencies, along with all the laws under color, all admiralty laws, and all lawyers.

Atheists are like some religious, they push their agenda on others. An atheist is an egotistical idiot, and materialistic fool, but the anti-Christ own the world.

The WO owns your house, your car, your land, even your "children." Fail to pay contribution or compensatory assessments (taxes) to them and they will take your home, children, and all. They take your children into bondage and brainwash them. They claim this is for their own good. Fail to pay taxes to the WO and they will discard you (and your family) like a sack of rotten potatoes. They make laws to imprison people, because prisoners are slaves. The WO agenda does not include the well-being of anyone, especially US "citizens," save themselves. They use us as medical guinea pigs. As you probably know, there are several chemical poisons masked in inoculations.

Why should we "save" this planet? The earth is not ours. We will not inherit the earth, nor will our offspring. As we go along with environmental laws, we succumb to the lies and controls of the WO. Yet, we do not want to assist in poisoning ourselves.

This Earth belongs to satan, and all her minions and devils. About 2000 years ago satan tried to give this Earth to Jesus, but Jesus had a much better home. This is satan's world, therefore, this is the WO's Earth. Why should we save this world?

When watching continuing debates between evolutionists and creationists to allow creation to be taught in school, one begins to realize that scientists are reaching the point where they, and/or their "conclusions" cannot be questioned. This indicates that someone is protecting a lie.

That which will kill in high doses, will harm in low doses.

That which will kill in small doses will harm in minute doses. That which can kill in a short time can harm when spread over a long time.

Beware of EME being forced on us 24 hours a day, 365 days a year, our whole lives. Beware of medicine forced on us and our offspring our whole lives.

This world is of satan, so Christians are really the aliens.

The WO, through big business and "laws," force their whims and wants on us such as fluorinated water, inoculations, prepared food and drink, medicines, EME, media, and sprayed chemicals. The big rub, we pay for it.

By now you have probably connected the WO to the anti-Christ. This is correct. The WO order has been around for thousands of years, and so has the anti-Christ.

Requirements To Regain Our Constitution

Not that the following list of deletions will ever be done, or even started, and probably not even thought about. These are things that must be completed before we can ever return to our Constitution and control of our rights.

1. Delete and prevent all insurance.
2. Delete and prevent all "laws" under color.
3. Delete and prevent ownership of land by any of our governments.
4. Delete and prevent IRS and any form of compensatory taxes.
5. Delete and prevent FBI, CIA, ATF, etc. authority to arrest or even carry weapons. Delete CIA authority in USA.
6. Delete and prevent confiscation of "property" by any government or their agencies.
7. Delete and prevent all taxes under color.
8. Delete and prevent all contracts (including government) not signed and entered into knowingly, willingly, intentionally, and voluntarily.
9. Delete and prevent admiralty laws within the United States of America.
10. Delete and prevent lies, twisted truths, and non-truth in advertising, politics, science, teachings, schools, medicine, and business.
11. Delete and prevent use of armed forces overseas, and maintain the size needed for protection of the USA, and create real Citizen militia.

12. Delete and prevent non-Constitutionally accepted amendments.

13. Delete and prevent brainwashing policies and tactics in media and education.

14. Delete and prevent medical propaganda and forced use of medicines.

We are again brainwashed into using satan's wishes. When we say we are proud of someone we are expressing and describing ourselves as arrogant, conceited, self-praised, self-esteemed, egotistical, boastful, overbearing, and self-idolizing. Instead of being "proud" we should express our "excitement" by being very happy for someone.

Anyone committing, or agreeing with, abortion is also accepting and concurring with murder. Those who are not against abortion may never be trusted for any truth, especially politicians and medical people. Once again: Anyone who motivates, inspires, or instigates abortion of the unborn, should be aborted.

It appears that the more "Scientists" learn about the universe and science the more they find out how wrong they are. Heaven forbid they ever admit before they "discover" something that they may be wrong. The real irony is that they will never know enough to be right. Some "Scientists" are such egotistical idiots.

We need to be masters of our past, rather than slaves in our future.

We need to be masters of our politicians, rather than slaves of our government.

Lies of our government and scientists must be continuously reinforced through rules, laws, and more lies. Good examples of this are the "holocaust" and "global warming." And, may the wrath of the world order fall on "doubters."

The most dangerous terrorists are in Washington, D.C. Yet, they are only puppets of the WO. We can say the WO are the most dangerous faction ever, but we do not know where they are.

In the spirit universe there is no time, distance, and velocity as on earth where we are transfixed on these dimensions, plus flesh.

The really lucky ones are those who are no longer in the flesh. Soon we will all be lucky.

One of the many, many ways the WO conducts is "culture conditioning." Since Americans are lazy, cowardly, greedy, gullible, and egotistical, culture conditioning (brainwashing) is an easy task.

How can we control gravity? What is gravity? Gravity is energy, specifically electromagnetic energy (EME). The universe is completely filled with EME of all frequencies and wavelengths. Despite what scientists tell you about gravity, they are either lying or very stupid, or both. Gravity does not come from mass. Mass only absorbs a little of the EME to create a difference of "pressure" between two mass objects.

Gravity is EME, but what is mass? Mass is only latent EME. Mass absorbs EME as latent energy. So, how can gravity be controlled? When an object is moving through space it is due to gravity, such as the orbits of planets and moons. If some of the EME in front of an object can be channeled around to the backside there would be less resistance up front and more push back aft. When one thinks about it, this could be an avalanche effect. The faster the object moves the more power available to move this object. Gravity does not "attract," it repels.

Mass can affect EME, and EME can affect mass. The mass of the sun causes light to bend. The mass of the earth also bends light. Magnetism also bends light. Some how this indicates that magnetism is also associated with mass.

Mass is EME, gravity is EME, magnetism is EME, and nuclear energy is EME. Electrons, protons, neutrons, and even some photons are constructed of latent EME.

The mass of a single photon (one half EME wavelength) is about $3.686E^{-48}$ gram. Doesn't this sort of combine the four forces? All energy is EME in one form or another. Thus, the manipulation of gravity is a real possibility. The control of gravity will be handled through some form of EME. If we could only figure this out; or learn it from aliens.

The only advances in science have been accomplished through either accidents or trail and error, not theory. Theories seem to appear after the fact, and even then most of them are wrong.

This world is filled with lies, it is overflowing with deceit, greed, and ego, and little is done or even cared about the opposite, it is condoned and perpetrated by "free" Americans. But, try to tell the truth and you will be persecuted, bullied, and truly "crucified" by the establishment and its minions. This is intrinsically and verifiably satan's earth.

It's too bad American heroes aren't teachers, or doctors, or nurses, or firefighters, but they are people in sports, politics, entertainment, and other minions of satan. This is a good indicator of our lack of intelligence. This admiration for false heroes is a very good indication of "culture conditioning" and "attitude engineering."

The spirit and the flesh are always against one another. The spirit is God serving, and the flesh is self serving. The spirit serves God and the flesh serves satan.

Lying scientists are trying to convince us that the earth has a fever. Maybe we should give it an enema, it would be placed in Washington, D.C.

There are many people, including Jews and Christians who worship a "god," but it is spelled with an "L,"—"*gold.*"

Man's "best friend" is also man's best friend spelled backwards.

There may be life on earth, but there is no sign of intelligence.

Scientists are educated idiots. Education is the training that helps make them egotistically saturated, thus becoming "idiots." The universe would give up more secrets if scientists would give up their egos.

The Constitution of the People is eroded through "political correctness."

The people do not want this erosion, until they are told they want it by the "politicians," and the WO.

There are many programs on radio, television, and the movies that are designed to reduce intelligence, and convince us the WO is the right way.
A few of these are:

gay stories	sports
tattoo stories	sex stories
news stories	sitcoms
entertainment programs	police stories
rap crap, hip hop slop	court stories
public broadcasting stations	movies
ads	"real life" shows
talk shows	contest shows
top model	TV preachers

When lawyers and politicians can't get what they want by twisting the truth, they twist the facts and emotions.

Remember, the new world order (NWO) is not really new. It has been in existence for well over 5,000 years. They are truly the old world order (OWO) coterie of masters that have dictated "laws" under color for many millennia (including the colorable laws in the present United States). We'll just continue to call it the world order (WO).

To crush something, do you pull it from the inside or push it from the outside? Gravity does not originate from mass. It is very hard to un-brainwash someone who has been trained all their lives. It seems like ego grows with "education." Scientists are a fine example. The higher the IQ you have, the easier you are to train, and wash your brain.

A star does not die, it just divides and regroups.

The more open your mind is, the more truth it can recognize.

How smart can scientists be? They believe gravity originates from mass. They believe gravity is a pull force. They believe orbits are elliptical. They believe in the big bang. They believe the evolution theory. They believe in time dilation. They believe in global warming. They believe the moon's gravity is 1/6 earth's gravity. They believe in space/time warp. It seems as though scientists believe the lies, but not the truth (just like the rest of us).

Like politicians, CEO's, insurance companies, banks, oil companies, telephone companies, etc., scientists believe in deceiving the laymen, the common, and the naive. They speak a lot, but they say nothing, they know nothing, but deceit.

Many scientists throughout the world belong to the EEII (educated egotistical intelligent idiots), an old association just now made up.

What a beautiful world this would be if we only knew the truth, or just a little bit of it. If the truth was a million miles long we only know just one micrometer of it. To know the truth, the whole truth, one would need an IQ of infinity.

Did you know that around ninety nine percent (99%) of "laws" (about 5 or 6 million) in the U.S. are unconstitutional? Of course, if there is no Constitution, nothing is unconstitutional.

The most cruel, inhuman, barbaric, vicious, diabolical, and evil terrorists that have ever existed on this planet earth are here now controlling many countries, including the United States. These terrorists are old, very old, they are the Kenites (WO).

The EEII insures that the common people are continuously astounded, amazed, flabbergasted, perplexed, and confused. The covert EEII membership is not exclusive to scientists, all disciplines (politics, religion, business, etc.) are favored with intelligent idiots. They are not really doing their jobs, they're doing their ego.

Who are the scientific idiots that say blowing up a comet or asteroid that is threatening earth would just make smaller pieces to hit the earth? Isn't that exactly what we want? Most small pieces would burn up in the atmosphere. Plus the fact that blowing something up makes pieces go in many different directions and trajectories.

Half of physics, half of astronomy, half of astrophysics, and all of cosmology are straight up lies, created by EEII and the old world order (WO). Although, most scientists just go along with the flow, being super brainwashed, egotistical, cowardly Americans.

If you think that Americans can't be brainwashed just remember the experiment the WO masters tried on Americans. They put out the information that the year 2000 was the beginning of the 21st millennium, and ninety nine percent (99%) of Americans (and others) believed this lie, along with the "doomsday" fear factor for many computers.

This is a short list of scientific lies:

carbon dating
Einstein's E=mc2
elliptical orbits
evolution
big bang

light speed limit
moon's gravity
no cold fusion
ozone layer hole
time dilation

fossil fuel	time/space warp
CO2 causes global warming	UFO excuses
light interference patterns	zero mass photon

There are many, many others too numerous to list. The majority of them are in theoretical physics, and especially cosmology.

It is illegal to bury or dump fluoride, even in a waste dump, yet our wonderful governments, in their "wisdom" are injecting fluoride (a very toxic chemical) into our drinking water. They don't suggest that we use it on our teeth, but drink it. That is like drinking suntan lotion to prevent getting burned. We could call it very stupid, but it's far worse than that, stupid implies innocents.

When our "leaders" want to promote and expedite (to railroad) a product or concocted idea they circumvent the suspicions by promoting a fear factor. The WO are by no means stupid, and by no means new. They have brainwashed over 99% of Americans, and no telling how many people in other countries.

It seems like scientists rival politicians in the liars category. Everything they "discover" in space is due to the "big bang." Everything happening on earth is a result of "global warming." If Americans believe this crap we deserve to be duped.

Global warming will happen, but CO2 has nothing to do with it. Helping it along are the thousands and thousands of "contrails" which are actually "chemtrails" distributed by the world order, mostly through military aircraft. But there are other causes such as natural energy absorption by mass (earth's mass). Eventually the earth will become like Jupiter, then a star. Jupiter will also become a star, long before Earth.

The History channel produced a series called "The Universe." Almost all of the information presented has been so contrived and twisted that it is actually straight lies. How can such educated people consider these theories? Oh, yes, brainwashing. The more intelligent one is the easier it is to be brainwashed. When universities teach theoretical lies, intelligent students become brainwashed, and pay for it themselves. If one would desire to get closer to the truth, reverse one's judgments, convictions, and conclusions. An open mind can contain an infinite amount of knowledge.

Before "global warming" can effect dangerous climate change, other disasters may occur that will annul any warming trends. If only one volcano creates a major eruption it may send the earth into global cooling, or even a mini ice age.

Fear Factor Farce Fecundity

Americans are basically cowards. We are so afraid to fight for constitutional rights we cower on the easy chair watching others be heroes. The WO knows this very well. Much of their strategy for control includes creation and installation of fear in their "subjects." There have been many fear factors for Americans, who pass it on to their heirs.

If one will notice, the WO continuously devises and contrives fear farces. A few of these are "global warming," "terrorism," and "comets." There are many, many more to keep us cowering in our "security." As we are in fear, then we are subject to more and more "laws" that scrape away our rights.

No rights, no freedom. No freedom, no security or safety.
The only bright side is that there are no rights left to erode.

Remember folks, we are paying much, much more taxes to the WO than the American Colonists gave to England, and they fought a war for it. The WO has burned our Constitution with contracts of the sea, and colorable laws.

We are a screwed people. We seem to love being screwed. Following is a short list of our "screwers" (not in any special order): governments, oil companies, auto companies, steel and aluminum companies, insurance companies, banks, construction companies, phone companies, drug companies, media (all), police, judges, lawyers, doctors, etc. All these are condoned and encouraged by our local, state, and federal governments, who are our biggest "screwers." There are many others, but it would take a life time to list. Our "elected" representatives, the ones we trust to prevent us from being screwed, are the biggest "screwers." They make the "laws" that protect big businesses rather than us lowly common people (us small "c" citizens). When you hear politicians talking you know they're lying. Why cannot we make these "laws" to prevent all business contributions to politicians? Because "laws" are made by politicians. There may not be an honest politician from the state level up, and maybe some below the state level.

The number of politicians running for an office should be limited, and each one funded equally through the tax system funds, other contributions prohibited.

One may start out as an honest politician, but soon gets sucked up in the conspiracies. The ones who try to remain honest get framed as criminals, or assassinated. The WO and its agencies (FBI, CIA, ATF, etc.) are very, very adroit and proficient at creating false evidence, connected to lies that are almost feasible, against someone, then convincing (brainwashing) the public to condone and justify these accusations, using the WO owned media.

We sure have false hope. Sometimes when we see an advertisement, we run down and buy the product hoping it will work. Ninety nine percent of the time the product is a farce. Yet, we'll return again and again to buy another false product. Wouldn't it be a wonderful world if all advertisements were true?

If our governments were truly interested in "saving" the earth they would stop destroying the rain forests, and stop spraying us with chemical contrails (chemtrails). All they want to save is their control over the people.

Remember, the most dangerous terrorists in the world are the US governments and their agencies.

If you give so others can win, you win. If you give so you win, you lose.

Something ain't right! We pay companies (auto, banks, charge card, insurance, phone, cable, etc.) for service, and they write the contracts that we get stuck with. In the real business world the ones who are paying are the ones who write the contracts. Not us Americans, we're stupid enough to let the companies control the contracts.

What's worse, the banks and credit card companies charge us an annual fee so we can pay them a monthly fee. We also pay the banks to profit from our money. This is basically because the banks, et al, compensate our "elected" politicians.

The truth in government is thousands of times more elusive than Big Foot, Nessie, Champ, UFO's, etc.

If you don't think Americans are stupid, just list who our idols are.

If A Tree Falls

There are answers to all questions, even some of the hard ones we always ask:

1. "If a tree falls in the woods, and there is no one around, will it make a sound?"
2. "Which came first, the chicken or the egg?"
3. "What does God look like?"
4. "When will the world end?"
5. "What is the meaning of life?"

Answers:

1. A falling tree will make an air vibration (sound), but not a noise.
2. The egg came first, it was laid by a bird that was not yet quite a chicken.
3. If you wish to know what God looks like, look at each other.
4. This world ends when we die.
5. We are here to choose between the Creator or His Creation.

Simple answers, but true.

The seventh month reveals the deception of Americans and others.

Americans must be bright, because our stupidity shines brilliantly.

When WWII began for Americans we started building thousands of airplanes, tanks, guns, uniforms, etc. The beginning of WWII ended the great depression. Have you thought about where the billions of dollars came from to fund the war? It wasn't there before the war. Americans were in a very deep "depression." The U.S. government performed magic. Or was it slight of hand. Yes, the great depression was created and orchestrated by our "wonderful," "elected" government (WO). Once again the American people were conditioned for the next round of lies.

If one thinks Americans aren't created and conditioned idiots, just count the auto accidents occurring each day. This partially stems from lack of responsibility because of auto insurance.

If the old world order can convince Americans that the year 2000 started the new millennium, then they can convince Americans of anything, especially global warming. It was a test, a gullibility test. Americans made a 100% grade (A+) on a stupidity test.

We are much more than just stupid and gullible. The way we lost our Constitution was being greedy, lazy, cowardly, stupid, and brainwashed.

When TV stations present specials on the universe they interview "scientists" (astrophysicists, astronomers, cosmologists, etc.) who talk like they know what's happening a billion miles away. They treat their assumptions as gospel truths. Such ego! Once again, we are educated, egotistical, intelligent idiots (EEII).

It is truly a wonderful thing that we cannot understand the weather. When we understand something we become familiar with it. When we are familiar with something we are not afraid of it. When we are not afraid of something we use it. When we use something we eventually control it. When we control something we eventually misuse it. Nature is truly good to us when she precludes our understanding of it.

There are many "conditionings" of Americans. One is leading us into the deadly sin of Pride. When any of ours does well we are "proud" of them. Pride is the first listed of the seven deadly sins. It is now common to see "Proud of" bumper stickers, and listen to "I'm so proud of you." Shouldn't we just be happy for them, rather than advertise that we have succumbed to the leadership of satan?

We fall smack into the pitfalls, performing all the 7 deadly sins, which are:

Pride, Greed, Envy, Gluttony, Lust, Anger, Sloth.

On UFO's:

Why do many government records and reports disappear?

Why are very lame excuses given by governments?

Why do UFO's show themselves?

Why do professional pilots get lambasted for reporting a UFO?

Why does the History channel show UFO programs? Are they creating lies like they do for global warming?

Why aren't photos of UFO's clear?

Why would UFO's have strobes or lights?

Why are UFO's so fast? How can they maneuver so sharply?

Why are there depictions of UFO's in ancient paintings?

Are UFO's American made, foreign made, alien made, or all three?

Are UFO's connected to the world order?

Why are we so ignorant of the universe?

There are some clues pointing to "visitors" from "outer space." A few of these are: Stonehinge, pyramids, Easter Island Moai, Nazca lines, crop circles, Area 51, UFO sitings, UFO crash sites (Roswell, etc.), "New" world order, Government cover ups, Tiahuanaco, Atlantis, Machu Picchu, Chichen Itza, and Fatima. Of course, the "New World Order," as you probably already know, is not new at all, but old, very, very old, more than 5,000 years old (maybe 10,000 years)

The Sphinx has water marks from rain storms which occurred more than 10,000 years ago. Of course, the Egyptians' pride would never allow them to admit that they did not build the Sphinx or the pyramids. Like many scientists, they ignore or twist scientific data, go for the wants and wishes, and manufacture evidence.

Cosmologists are like the Egyptians. They take all the evidence the universe shows them and misinterpret it to fit their own agenda.

As books are misinterpreted when rewritten, and the meaning of words change, true history is lost and false history is then enjoyed by all.

You own your own life, but when you listen to your peers, they will own you. If you dress as a whore you will attract a pimp. When you dress as a princess you will attract a prince.

Why are we so stupid that we tolerate abuse by law makers, law enforcers, and lawyers? We let false law rule our rights and destroy our freedom. The so-called "public policy" is really government control.

Politically there is a "left" and a "right." *They call themselves "right" rather than "almost right." They call themselves "left" rather than "wrong."*

Liars always claim they are telling the truth, and accuse others of lying.

History is created through lies. Then again, so are laws, science, physics, cosmology, religion, news, promises, ads, etc.

Homosexuality is a far worse affliction than homophobia.

Our government is in league with satan. They lust for total control. Their lies create excuses to further this control.

Fascism advocates dictatorship through merging of state and business. Ain't that what we kind of have now?

Education will rule ignorance, but evil rules the world.

God gives us free will, but our government takes it away.

Dictatorship, Totalitarian, Theocracy, Monarchy, Aristocracy, and Plutocracy (but no Democracy or Republic) are rolled into one to create the United States Government. We suffer from communism, fascism, socialism, and terrorism through "our" government.

Look at the signs and symptoms: no rights in court, very oppressive government, trend to poverty, police state, anti-Christian, many, many laws under color, and out of control deception by government.

The reason conspiracies are so extensive and unbounded is because people do not believe contrivance, stratagem, and secret societies exist (gullibility).

Our Constitution was void when we agreed to cease fire and submit to King's rule. We do not even come under the Magna Carta, because the "King" uses admiralty law (maritime law). This is depicted by the yellow fringe on the sides of the "American" flag.

To gain the Constitution we never had we must destroy the political and economic capacities of all lawyers, judges, Kenites, and England.

The entire world together can not defeat the U.S., our defeat came through inside invasions from the Kenites.

Even though it seems incredulous and unbelievable, we are under the British rule of Admiralty. The British are still under Roman rule. Rome is under Vatican rule. The Vatican is under the Jewish rule. The Jews are under Kenite rule. The Kenites are under the descendants of Cain rule. Cain's descendants are under satan's rule. This is the tree of knowledge.

The "monies" from income tax against Americans goes to the Crown of England (and its controllers). Not one penny of income tax goes to fixing, helping, or running the United States of America the Twelfth. Yep! We sure do have some "wonderful" politicians.

Americans are so brainwashed with lies we not only cannot recognize the truth, we vehemently deny it. If Americans new the truth we would be too lazy, too greedy, or too cowardly to do anything to correct it.

As we have discussed before, the "New World Order" is not really new. It is thousands of years old. There is not a new government conspiracy to control US or world Citizens. We are already controlled, and we are no longer Citizens, we are citizens. Our names are written and printed in capital letters, meaning we are of corporate persons (artificial or fictitious). Therefore, being fictitious we are subject to the laws of Roman rule and the rule of Admiralty (merchant law). The insignia of Roman, admiralty, and merchant law is a flag, ensign, or banner with three sides lined with a yellow fringe. Roman law rules England, England rules with admiralty and merchant law, admiralty law rules the United States. The United States of America the Twelfth did not truly exist. Our heroes deceived us when they claimed that we "won" the Revolutionary War. Instead "we" agreed to stop fighting if we comply with admiralty law. A few of our lawmakers disagreed with this decision and created the 13th Amendment. During the war of 1812 the "British" burned our capital and destroyed all records of certain amendments.

The American civil war had absolutely nothing to do with slavery. It was a police action to retrieve Congressmen that refused to accept the British rule of admiralty imposed on the United States of America the twelfth. These laws continue to create the United States (the corporation) and all its citizens.

The conspiracy does not end there. The greatest part of the conspiracy is deceiving Americans into believing they have a Constitution.

Since we really do not have a Constitution why do we have a supreme court, or elections, constitutional objections, or limited terms of office, or voting machines, how can federal law override state law, etc., etc. Deception is the number one conspiracy.

The controllers of this world are either alien, human, satan, or any combination of the above.

The History channel, et alii, are like politicians. About ninety percent (90%) of presentations are lies, nine percent (9%) wrong, nine tenths of one percent (.9%) accidental truth, and one tenth of one percent (.1%) are arguable.

The scientists they interview are definitely educated egotistical intelligent idiots (EEII). How can these "intelligent" people take data from cosmological instruments and apply it to the wrong theories? Evidently, it's applied, not scientifically, but egotistically.

There are many scientists that know global warming is another world order scheme for more control. The WO has many methods of creating control. A few are: Taxes, laws, scare tactics, media, and education. There are more, but we get the hint. The WO controls the people that control us. We are very controllable. The "New World Order" title is itself a lie. It is not "New" it has been around for a long, long time (over 10,000 years).

And, "World" may not be the only place the WO controls. They have been deceiving us in lies about the "holocaust," "Hitler," "global warming," UFO's, and many, many other subjects (almost everything we "know") for hundreds, or even thousands of years.

To oppose these lies is a career ending action. To expose even a small part of the grand conspiracy can be much more than just career ending, it can probably be life threatening, or ending. At minimum, it would be prison time due to the actions of our beloved federal agencies (FBI, BATF, CIA, etc.). Prison time is also given to those who try to claim their constitutional rights, not knowing their rights are on Pluto (the ex-planet).

Money, as plastic or fiat, created by the world order is in our hands. We have been brainwashed with lust and greed that is now in our heads (foreheads). We bend to the will of the beast (world order) and satan, she is the wind that bends our will.

Don't you think there is something "funny" (weird) with our tax laws when the most pay the least and least pay the most? Then our wonderful government, that is all lie, throws us in jail because they accuse us of lying.

Who and what is the world order?

If human: their stratagem, design, scheme, and masterminding has been passed on from generation to generation, retaining the goal of their ancestors.

If alien: their goals may or may not be passed on depending on the changing human race, and also on how long the aliens live, and what their motives may be.

If satan: then her goal has been around for eternity. Also, she has the intelligence to devise, concoct, and contrive the lies and deceits needed to fool us, she knows us very, very well, she and her minions are "familiar spirits."

Humans are "intelligent?" But, we are not trustworthy, not faithful, nor truthful, we are proud, greedy, lusting, envious, lazy, and easy to anger. These are all traits that satan has whispered to us. We are constructed from egos, lies, and stupidity.

Politically left is freedomly wrong.

An idea is not a theory. Inventions do not normally come from theories. Inventions come from ideas, trial and error, hard work, and by accident.

Theories evolve after the fact. And then, most are wrong, some on purpose.

A few theories that are wrong are: Big bang, carbon dating, earth creation, electron superposition, elliptical orbits, evolution, expanding universe, "fossil" fuel, fusion powers stars, CO_2 causes global warming, gravity sucks, light interference, light speed limit, moon's gravity, cold fusion, ozone hole, rapture, time dilation.

These fabricated, fraudulent, fantastic theories are brought to you compliments of our wonderful world order.

You know you are under a police state when:

You are thrown in jail for violating a "law" that is under color (not real).
You are thrown in jail at the whim of a judge.
A judge will not allow you to defend yourself.
A judge will not tell you what kind of court you are being tried in.
The FBI can shoot and kill a mother while holding her baby in her own home.
Cops break down your door, for any reason.
You are thrown in jail for not having insurance.
You are thrown in jail for trying to exercise your "rights."
A prosecuting lawyer can decide whether you are an adult or child.
Cops can confiscate your "property."
You are thrown in jail for growing a natural plant.
You are thrown in jail for walking across the street.
You are thrown in jail standing around.
You are thrown in jail for not having a home.
You are thrown in jail for having a pet.
You are thrown in jail just because someone accused you.
You are thrown in jail for killing a chicken.
They take away your offspring because you are sick.
They take away your offspring because someone accused you.
They take away your offspring because you are poor.
They throw you in jail for not paying tribute to the world order.
They encourage abortion, but throw you in jail for killing a dangerous animal.
They throw you in jail for protecting yourself or your family.
They will persecute you for not believing in the "holocaust."
And soon they will throw you in jail for not believing in global warming.

The problem with stupid people is that they judge others by themselves.

Most of the "laws" that jail citizens are **laws under color**, and are policies of the world order, not the public. There is a great rainbow over the U.S.

Ignorance is bliss, but it doesn't leave anything to think about.

Ignorance is bliss, except for maintaining freedom.

Ignorance is curable until it reaches stupidity.

A "cover up" starts at the top and descends. If the UFO cartel can control top politicians and military then they control everyone. It is not hard for records to vanish or appear when the ones that keep records are the same that support cover-ups.

Our government and its politicians are run by the WO global business plan.

The US Government is not a nonprofit organization; it is a for profit corporation.

BOOK 3

ARTICLES
Religious Themes and Aphorisms

Introduction

The truth is put forth with humble approach, and minimum of conversation. If any of these writings show wisdom then they are not the author's, but of our Lord God and His Holy Spirits.

Please read this with trust in things new, and a little patience in things old. Although, some words seem to convey very new ideas, remember always that there is nothing new under the sun.

Any science discussion explains God's universe, and its composition. God is endless, eternal, and infinite, and so is the universe, because it is of God. The earth was created from the universe, which has no beginning, nor end.

This is to my sons, and to my beautiful grandchildren, who I wish I could guide to God.

Please read and retain these writings and the scriptures. Learn the scriptures of old, not the English version, ASAP. Believe that Christ Jesus is real, I know, for he touched me several times. And if Jesus is real then his Father is real, and if his Father is real then also are his Blessings, and eternal life in ecstasy. There is no way to explain the ecstasy of the touch of our Lord God. There are no words in any tongue to express that joy. This happiness is beyond our wildest wishes.

With all my spirit and soul I pray our Lord continues to bless you all. Do not forsake him, and he will never withhold his favors from you. he often does not like the things we do, but he loves his children more than eternity.

Remember, all things are miracles.

God is our Creator, his son Christ Jesus our Savior, the Holy Spirit our Teacher, the Angels our Protectors, and the Saints our Examples.

Thank you Father for our daily bread, our guidance to you, and all your Mercy, and for your son Christ Jesus, our Savior. Heal us, O Lord, so we may be with you now and forever.

The Bible has many interpretations, and many misinterpretations. The Bible is constructed in seven levels, of which the best and gifted can only know up to the fifth level. An example of the third level is an explanation of Rev. 10.

Revelation 10

"And I saw another mighty angel come down from heaven, clothed with a cloud: and a rainbow upon his head, and his face as it were the sun, and his feet as pillars of fire:

2   And he had in his hand a little book open: and he set his right foot upon the sea, and left on the earth,

3   And cried with a loud voice, as a lion roarth: and when he had cried, seven thunders uttered their voices.

4   And when the seven thunders had uttered their voices, I was about to write: and I heard a voice from heaven saying to me, Seal up those things which the seven thunders uttered, and write them not.

5   And the angel which I saw stand upon the sea and upon the earth lifted up his hand to heaven.

6   And sware by him that liveth forever and ever, who created heaven, and the things that therein are, and the earth, and the things therein are, and the sea, and the things which are therein, that there should be time no longer:

7   But in the days of the voice of the seventh angel, when he shall begin to sound, the mystery of God should be finished, as he hath declared to his servants the prophets.

8   And the voice which I heard from heaven spake unto me again, and said, Go take the little book which is open in the hand of the angel which standeth upon the sea and the earth.

9   And I went unto the angel, and said unto him, Give me the little book. And he said unto me, Take, eat it up; and it shall make thy belly bitter, but it shall be in thy mouth sweet as honey.

10   And I took the little book out of the angel's hand, and ate it up; and it was in my mouth sweet as honey: and as soon as I had eaten it, my belly was bitter.

11   And he said to me, Thou must prophesy again before many peoples, and nations, and tongues, and kings."

The cloud is an obstruction and occultation of the true cloths of the one offering the little book. and depicts someone who is not to be trusted. The rainbow over his head depicts someone under color, using the laws of the sea on land as real and true laws. The face as the sun is to blind one to the truth, as would the false messiah. The feet as pillars of fire depicts the lake of fire supporting the deceit presented.

The little open book is admiralty, maritime (Roman) law with the "benefits" replacing freedoms. One foot on land and one on the sea depicts the Roman law of the sea that has been expanded to encompass and usurp the constitutional laws of the lands, and nations, as done by the "Elite" through a few evil lawyers and the false lights.

By the seven thunders being done, it is too late to correct the evil deceit we have already swallowed, along with the unrevealed contracts, when they are revealed how bitter they truly are, but too late, much too late. The earth belongs to satan, there is no truth on earth.

Remember, we may not be able to beat satan, but we can beat the imps who serve him, by using their own sins against them. God has given us power to vanquish the evil inside us, and in others, although we appear to pride ourselves in evil. Why are we so proud of our ungodliness? Where will be our pride in the lake of fire?

My flesh loves the Creation, but my spirit loves the Creator. God gave us a choice, we can worship the Creator, or worship his creation.

Does God judge us by how much we love him or by how much he loves us? What he does for us, or what we do for him?

Love is self perpetuating. The more people you love, the more love you have for each.

Although, I pray that you find your way to the Lord, because I respect your soul.

I thought I was lucky when I saw a man with no shoes. He said he was lucky because he saw a man who lost his feet. In turn this man said he too was lucky for he saw a man who lost his legs. This legless man also believed he was lucky, for he saw a man who lost his arms and legs. Again, this man with no limbs said he was lucky because he saw a man who lost his arms, legs, and eyes. And this lonely blind man said he was lucky because he knew people who were very, very rich, but have lost their souls.

Bless Those

Bless those who threaten me, for they test my courage,
Bless those who honor me, for they test my humbleness,
Bless those who hate me, for they test my love,
Bless those who smite me, for they allow me to turn the other cheek,
Bless those who aggravate me, for they test my patience,
Bless those who chain me, for they test my faith,
Bless those who use me, for they test my charity,
Bless those who love me, for they test my weakness,

Bless those who abuse me, for they test my forgiveness,
Bless those who fear me, for they test my compassion,
Bless those who insult me, for they remind me of what I am,
Bless those who misjudge me, for they test my wisdom,
Bless those who curse me, for they test my composure,
Bless those who defy me, for they test my peace,

Faith can move mountains, yet it alone cannot get you into Heaven. All things that will get you into Heaven hang on the Love for our Lord.

Why do we plan ahead? We plan for the weekend, for a vacation, for a retirement, and we do not know when God shall require our souls. As we have wealth, it will be gathered by others to waste upon pleasures of flesh and familiar spirits. Are we not wise enough to earn and save what we may take with us, and no one can steal?

We must earn love and faith in our Lord, acts and deeds done in His name, in His blessings. We shall be not ignorant of our true wealth and the future of our souls. We try to do unto others as our Lord has commanded.

We have four beings within us. They unite at our conception, and are in the likeness of the Seven Spirits of our Holy Father; they are our body, mind, soul, and spirit. Our spirit may reach out to infinity and understand, but our minds cannot. The human mind is not designed to comprehend infinity, neither in time, distance or velocity. We do not truly understand what we cannot observe.

Our mind can reach far in distance and time, but not to infinity or eternity. Everything we know has a beginning and an end, including ourselves. This is one reason Science is determined to convince us of the "big bang."

The universe is of God, therefore it is eternal. Our spirit is eternal, thus it is able to know infinity. Our brain and body are one, our mind and body are separate, yet one. A soul can be "locked up," but the spirit may still wander. The only control you have over your soul and spirit are your sins, or your love for God. As our mind and body dies our soul and spirit returns to our Father, our Creator.

The immortality of our souls depend upon immortality of our love for the Creator, his son, Christ Jesus, and the Holy Spirit.

I know all things in the universe will fail me Lord, either by power or by longevity, save your Word. *The evil shall perish as they know they are not with God. Their anguish and regret shall sear their souls from within.* This is the worst hell that can be suffered. All the pain and torment cannot even level with the loss of our Lord's love and comfort. Our Lord will return to us as we have given to him and to others. If you love our Lord you will be true unto him, and all things will hang from this. The love of God holds all righteousness. We must follow the teaching of our savior Christ Jesus, he is the way. And, the way is through him to our Lord and God.

Agnostics and atheists believe only in themselves, knowing they are of the universe which is not God's. They are ignorant of the Word. What one is ignorant of one despises, what one is afraid of one curses. "… but fools despise wisdom and instruction."—Proverbs 1:7. Agnostics and Atheists refuse to learn the Word. This is much more than ignorance, it is stupidity. Ignorance can be cured by study, but stupidity is fatal. Most atheists are satan saturated babbling fools.

As the sons of Cain appear in government, the first "year" they call the people Master. The second "year" they call the people friends. The third "year" they call the people slaves. Our servants have become our masters. Beware of those who call themselves public servants, and are not. Beware those who call themselves ministers, and are not. Beware those who call themselves evangelists, and are not. Beware those who call themselves knowledgeable, and are not. Beware those who call themselves righteous, and are not. Beware those who call themselves Christians, and are not. Beware those who call themselves Jew, and are not. Those who are truly not of Judah, nor from Israel, are the children of Cain, thus children of satan. Cain is no where to be found in the lineage (genealogy) of Adam to Christ Jesus, because Cain is of satan.

If mass can become energy, and energy become mass, then mass and energy are the same. If dust can become man, and man become dust, then dust and man are the same. Life does not come from nothing, it comes from God. Our lives come from God and shall return to God, but our bodies return to dust, as our souls and spirits return to our Creator. And, our Creator is eternal and infinite.

My flesh is weak, I raise my bow and place a trembling hand on the string and lo, His hand guides mine, the arrow is flung forever straight, forever.

The birth of a child is clearly a miracle, but a much greater miracle is the conception of a being. A child and its mother endured agony and pains in birth, but the egg and sperm endure greater troubles and trails before conception. The holy conception of Jesus was a great miracle, although all conceptions are miracles, even more so than birth.

At conception our Lord "assigns" and sends an awaiting soul to unite with the newly conceived child. This child, even though is only two cells, has a body, a mind, a soul, and a spirit. As this body and mind dies it returns to dust, the soul and spirit return to our Lord and Creator.

Conception is a blessing and birth is a curse (Gen. 3:16-19). This curse was brought from Adam and Eve's disobedience. The life of the flesh begins at conception, not birth.

Everything that moves is of our Lord's. All creation is not still.

Unless you consider God's mercy or wrath luck, then there is no such thing as luck. God controls all the odds.

What is the truth? The Father, the Son, and the Holy Spirit is the only truth, and no other.

If you have enough love to walk with the Lord, then the strength, courage, and wisdom will emerge.

Lord, even for all the unrighteous, stupid, and sinful things I have done, you have still flown me when I fell, floated me when I sank, and stood me when I stumbled. Lord, I give thanks for your mercy.

In the time of Jesus the cross was used very extensively to execute people. Many Priests, Rabbis, and laymen were put to death through crucifixion by the Romans. The cross became a symbol of evil, torture, and death. The cross was especially hated and feared by non-Romans. Even many Romans were leery of this type of torture; yet, they enjoyed the "games" and the entertainment.

Upon the crucifixion of Jesus the symbol of evil, torture and death became the symbol of life, love and forgiveness. Many Christians readily accepted death in the same form as the Messiah.

If you do not have love for our Lord you will have little chance of gaining his Heaven. All things that will gain you Heaven (ie. trust, faith, courage, strength, wisdom, righteous works, obedience, etc.) hang on the love for our Lord.

To have love for our Lord we must understand what he has done for us, and what he is still doing, and how much he loves us. He will give you space, plenty of space, to repent. Our Lord grieves to lose one soul. But, the choice is ours. He gave us freedom of choice, to love him or to love his Creation. To understand our Lord you must study the scriptures. The English version of the Bible lacks much, and has many misinterpretations, study the scriptures.

Our Lord tests us. That is why we're here in flesh. All but a few souls must go through flesh, and thus tests. Some souls are too righteous, and need but little testing. Others are tested for a hundred years, and still fail.

Plant a seed, then water it a little. If it grows then attend it. If its branches grow too wily, trim them gently. Feed it sunshine, shade it from moonshine. Tend it with love and truth, a soul is a very precious being.

Do not look at others as human beings, but as precious souls of God's creation, which he is pleased in. When you plant a seed you are not saving a person's life, you are saving a soul for eternity.

Judge not the works of others, but influence their actions for the future.

Do not worry if someone is good or bad, you will not be judged by the actions of others.

We do not grow plants, we only tend to them; only the Lord can grow a stem.

Those who are egocentric and supercilious scoff at a being that should be greater than they.

The only thing we should be proud of is that we are not proud of anything.

The gifts of satan are all that one wishes for; "good luck," riches, power, control, to be praised, to be envied, to be respected, to be served, to be revered. These gifts of satan are given to flesh. These gifts are also given by our Lord, in Heaven to righteous souls. Which gifts do you believe will last the longest?

Those who do not love and worship our Lord, worship the gifts of satan.

Those who think themselves wise, are foolish indeed.

The foolish earnestly love the creation greater than the Creator.

If a billion years is swiftly passed in eternity, what is only a lifetime?

Those whose pride is inadvertently revealed, lose their credibility.

Our sins can also be things that we do not do.

Such as we do not:

> Learn God's Word
> Love God
> Love Christ Jesus
> Help the poor, Visit the sick
> Visit the imprisoned
> Cloth the naked
> House the homeless
> Teach of God and His Son

A teacher whose credibility is diminished may as well become a carpenter.

It is very common to be ignorant, but very stupid to wish to remain ignorant.

Those who do not believe in Christ Jesus will believe in almost anything else. Those who do not believe in God, will believe in everything else.

The arc of the Covenant is God's:

> direction to us
> word to us
> grace to us
> covenant to us
> hope to us
> belief to us
> love to us
> instructions to us
> forewarning to us
> prophecy to us
> everlasting life for us.

Thoughts of the flesh have limits of the flesh. Thoughts of the spirit have limits of the spirit.

Love of the flesh leads to death. Love of the spirit leads to life.

The flesh has but a short life. The spirit can be everlasting.

The flesh knows limits. The spirit knows infinity.

Always remember, the tree of knowledge is satan our enemy, and the tree of Life is Christ our Lord.

To know all things other than God is knowledge. To know God is wisdom.

If you do not laugh when you win, you will not cry when you lose.

If you do not laugh when you live, you will not cry when you die.

When you read, study, and compare the English version of the Bible to the scriptures in the older languages you will find that the English version cannot be used to find the hidden messages. There are some acrostics and hidden messages in the Masoretic Torah and no where else. To learn some of these hidden messages and the intricate methods of presentation, along with the prophesies that are slowly unraveling, you cannot but know that only by supernatural means it was written. The more you study the Bible the more you understand what has happened and what will happen; but, do not be carried away in interpretation.

To be a true Christian is not a religion, it is a reality.

To be friends with the world is to have animosity toward God.

How can we curse our Lord's Creation simply because it does not conform to our wishes?

How can we pray to, and praise our Lord, then out of the same mouth, and often even the same breath, curse his Creation?

We defile ourselves, not from anything of our bodies, except the mouth and the mind; and, the greater part of the time they are not linked.

We have a very short life to prove to our Lord that we are worthy to be accepted by him. Even if we were worthy, we should fear that we are not. Never be above humble and believe you are always worthy of Heaven. Those who believe that they are worthy of Heaven are too egotistical to be. We would to be worthy of our Lord, but we will never make it no matter how hard we try. No one is worthy of our Lord's love, and his Heaven. But, our Lord is very forgiving, merciful, and loving to his children, and especially to those who love him. His Word heals our sick souls, and makes us immortal.

We are the dust that scatters with the wind. We will never return again as flesh. Our short life in flesh will determine the eternity of our spirits. *Our souls are destined by whatever our flesh has desired.*

We must help our Lord help us. He gave us the equipment to maintain our lives in honorable and righteous means. We must use his gifts to us to honor him for the Love he is giving to us. Our Lord is offering us the greatest gift that any can receive, the eternal life with him in ecstasy. And all he is asking is that we return his Love for a short time in this life he has graced us with.

There is nothing that our Lord cannot do, except evil.

If someone gave you fishing gear and taught you how to fish, do you think that he would wish that you should keep asking for his fish?

If someone gave you a car and taught you how to drive, do you think that he would wish that you should keep asking for a ride?

If someone gave you seed and taught you how to tend them, do you think that he would wish that you should keep cutting his flowers?

This is what we do to our Lord when we keep asking for things that we can handle ourselves. Our Lord has blessed us with many, many gifts, and tools to make our way through this short life, and we still insist that he does our work for us.

We are backwards in our asking. We ask our Lord to handle something for us and we promise to do something for him. This is backwards, we should do for our Lord all the time, and then ask for what we cannot handle. Although, if we love our Lord all the time we would need to ask for little, except maybe guidance and approval for our work, and for forgiveness.

The rich or the poor, the master or the slave, the dark brown or the light brown, the Christian or the Jew, the young or the old, shall all meet each in our Lord's face, and he will judge. Then, only the righteous souls will be with him.

Even so, our Lord meets our needs according to his plan. Prayer is still the greatest power on earth.

Guide us, help us, Lord according to your Mercy, that we may know your hand is upon us, and your foot upon our enemies.

Remember, the desires of the evil do perish.

Our Lord Jesus is the Light our God created in Gen. 1:3. Jesus is the Light and the Day; satan is the dark and the night.

Lord, your enemies are my enemies, for they have forsaken your Word.

Lord, let our enemies confuse themselves and speak at each other.

Those who whore after idols have souls, otherwise they are as the idols they desire; having no speech, no hearing, no smell, no touch, no walk, no sight, no heart, and no spirit. These idols give no mercy, no forgiveness, no love, and no blessings. They create nothing, nothing but death (money can be an idol).

The most comforting thing in the world, is faith in our Lord. The most rewarding thing in the world, is trust in our Lord. The most powerful thing on earth, is prayer to our Lord, it involves the most powerful in Creation.

As our Lord is our Father and we his children, when we go wrong he chastises us. It is not for his pleasure, rather for our profit. All things our Lord, does for us, or allows to happen, is for our profit.

Infinity describes distance, eternity describes time. Time symbols are man made. Soon all the time shall end, there will be no time, only eternity; then, eternity will describe us in heaven or hell.

All strength, courage, wisdom, and love come from the Lord.

Lord, you know our thoughts, our deeds, our feelings, our agonies, you know them all, our spirit, our soul, our past, our future, from before time to unto our sleep, and beyond; us you know.

Lord, your Mercies out number the drops in the seas, or the grains in the sand.

Those who need chastising need it not done from the righteous, because their own tongue shall put them down.

Never condemn someone for what they are. Pray for them to find our Lord. Prayer has much, much greater power than criticism. The only thing larger than the universe is creation, and prayer has more power than either.

The evil satan cannot physically touch you, but he can ratchet jaw in your "ear" and lead you away. He can also lead others to physically touch you, or harass you, or persecute you, but only as you and the Lord allow. It is quite a privilege to be persecuted because you love our Lord God and his son Christ Jesus.

We are tempted by the voice of our own flesh long before satan's whispers.

The love of our Lord is not religion, but reality.

My flesh loves the creation, but my spirit loves the Creator.

As humility fades, hope wanes, faith turns away, love wanders, trust perishes, patience disappears, self control is lost, then a fool is born, and a soul dies.

If I believe that I am not a fool, I have deceived myself. We can deceive others, and even ourselves, but never our Lord, he knows our hearts.

Fools destroy themselves, they need no help. The greedy are slain by their own prosperity.

Flesh is natural, eternity is supernatural. The flesh shall not see eternity because it shall change back to dust (as it began to do the day we are born).

Dandruff is to remind us what we shall return to.

We are natural, our souls are supernatural. Our naturalness sentences our natural bodies to return to dust. Our supernatural souls may gain eternal life, as judged by our Lord.

Miracles are supernatural to us, but are natural to our Lord. We are all miracles. The universe is a miracle, a Creation of our Lord God.

Religion does not save, change, or give a soul eternal life, reality does. One receives our Lord's grace when one realizes Jesus is real, therefore God is real, and then learns the Word of God.

The diversities in opinions of the scriptures are derived through our material desires, not our supernatural love. Even though we are natural our entire lives are guided in supernatural directions, to our conception to our flesh death.

If we would listen to our spirits we would love our Lord, and his holy son Christ Jesus, our Savior. If we listen to our flesh we will love Jezebel and satan. Jesus will give us eternal joy, satan will give us eternal death.

Religions that reject Jesus as our Lord's Christ also accept satan and his ways. Those who do not follow Christ are lead by satan. Only those who love our Lord have a buckler against satan.

Thus, satan is always calling us toward him and his ways. He wants us all, especially those who love Jesus. He whispers in our ears, he tempts us with flesh and desires. He is like a beautiful colored asp begging to be handled. But, satan's bite is silent, pleasant and deadly, eternal death is his poison.

There are many people and factions that insist the Church modernize, and come up to "reality" in the world. The Churches should never yield or deliver up to satan's world. People should return to the Holy Church, rather than the Church move forward to appease society. The love of God does not change, or modernize. God's truth is far, far ahead of society.

All things on this earth, except the love of our Lord are not important at all. It matters not what we have or what happens to us, save to love the Lord. The ones that truly have the spirit within do not search or desire the materials of this world.

Technology is of familiar spirits.
If deeds from the evil do not appear cruel, it is only an illusion.
People do not know themselves, so they despise others.
We do not understand the miracles of God because we do not know God's Laws.

All we know as truth is temporary. Only God's truth is forever. Man has no truth in him. If we knew God's Truth we would not understand it, we cannot even understand infinity, much less eternity. The Word of God was written through the Prophets by the Holy Spirit. The truth in God's Word can only be understood spiritually, and as God wills. We interpret God's Word through flesh, but we will only know it as spirits.

God's Word is written to you, do not let anyone rob you of it. Christianity is the only faith that completely changes one as they turn to it. Christianity changes one's attitude, one's eyes, one's mouth, one's ears, one's work, one's play, one's grief, one's love, and one's soul.

We may suffer from either well doing, or evil doing; either is a blessing if it is because of the love for God.

Thank you Lord for your blessings of love and mercy, and of the children you love. Lord please give us the wisdom, knowledge, and courage to be with you now and forever.

If you can love God while in the flesh, you will surely love him in the spirit. If the fallen angels could not love God in their spirits, knowing him face to face, how can they possibly love him when they are on earth?

Our Savior, Jesus Christ, turned the terrifying, hated and evil cross, the symbol of death of the Romans, into the loved and glorious symbol of everlasting life through his Crucifixion.

Those whose god is their own face, wealth, and sex, have pride in evil.

Pride contains arrogant, supercilious, haughty, self-esteem, and self indulgence.

Christ will also make you free from pride, and the bonds of smoking, alcohol, lust, and greed.

As satan is real, the technology of the past, now, and future is his, even 100,000 years ago. Even the sphinx and the pyramids are his. There is nothing new under the sun; satan is as real as sin.

One can live in truth, be saved, and obtain Heaven and God's blessings through the New Testament, but not by the Old Testament. One cannot comply with over 700 laws of the Old Testament and the Talmud. And, as one breaks any law of God one might as well break them all. This is also true of the Ten Commandments. We can comply with the teachings of our savior, Christ Jesus.

An analogy of the Testaments is: the Old Testament is sort of like an exploded view of the parts of the spirit and soul, and the New Testament is the owner's operating manual. The two Testaments are also the record of the genealogy of Jesus, our Savior, and they are God's letter of instruction and knowledge to each of us. For without it we are stupid.

Today, the crimes created through the WO make Hitler's crimes look like they're done by a three year old. A few of the people, that the hated Hitler persecuted, are now persecuting Christians, and destroying our American Constitution and Rights. Remember, the Jews who are not true Jews, but seeds of the serpent, are always creating evil, even in the best intended people. Like a true Christian a true Jew loves our Lord God no matter what anyone else does, or loves. The true Jew and the true Christian Gods are the same God. The false Jews (Kenites) are the sons of Cain, the children of satan, they have no truth in them.

Earth belongs to satan. Heaven belongs to Christ Jesus.

We are like clouds. We form from the dust and water. We take many shapes. We move about, sometimes silent, sometimes with great noise and light. We often shutout the sun, stars, or moon. We gather together and create crowds of dark clouds to frighten earthlings. We grow to great heights, filled with fury. We release ourselves into a great river, and flood the helpless, and the poor, and the weary. Thus, as we came, we vanish into nothing, never to be seen again, and others shall repeat our life of sin.

If you do not understand God's Word, you do not understand how to quicken your spirit, therefore you greatly err in the flesh.

Velocity is unity. The mathematical formula for velocity is distance divided by time. Infinity represents distance, eternity time. Infinity and eternity also represent forever. Therefore, two things that equal the same thing also equal each other, and anything divided by itself is one. Infinity divided by eternity is one. One is unity. Thus, perpetual motion is also unity. Our Lord God is everything, and also Unity.

Our Lord is everywhere, in everything. Should he forget us for one yoctosecond we would immediately perish, and cease to exist.

As we cannot perceive zero, even less eternity, we cannot understand the "abstract" and things not of earthly values, therefore we cannot comprehend true wisdom because all wisdom comes from God, it is God.

Remember, we suffer from two things, well doing, or evil doing.

The mercy of our enemy is deceit.

Hunger is a poor test of taste.

A bloated belly despises cake, but to the hungry every bitter nut is sweet.

Do not flatter a friend, least he swells his head.

You can see a friend's heart through your own, and see your own through his.

Witchcraft, the control of another, is not limited to voodoo, witches, warlocks, and weirdoes. Witchcraft is the control of others. Control of other persons can be by ones who love them. This is a slight form of witchcraft. To control another one needs not to use ingredients of crushed bat feet, eel's eyes, etc., but only the personality of desire. There are many evil traits associated with witchcraft. Included in these are uncleanness, lasciviousness, idolatry, hatred, emulations, wrath, strife, heresies, jealousy, and even "love." The greatest example of witchcraft is our overpowering politicians, and the people that control them.

God and gravity! What do God and gravity have in common? They are one and the same. Both are everywhere, forever, and in total control of the universe. Gravity is of God, God creates and maintains gravity. Gravity employees the smallest particles to create the largest force. Gravity also uses these "infinitely" small particles to create the largest bodies in the universe. Gravity, through God, is the origin of all forces in the universe. Gravity, through God, maintains life, causes death, is no respecter of persons, cannot be controlled, and is always present, always the same, and will be forever. Gravity rules the universe, yet God rules something greater—Creation.

To be a true Christian is to be far from religion. Religion is of man's ideas and traditions. True Christianity is not tradition or fiction, but fact. When one truly receives Christ Jesus, one is totally changed in appearance, in desires, in attitudes, and in spirit. There is no other religion that enhances a soul so righteously, none.

Know your enemy, but do not reveal your knowledge. Know satan's tricks.

Know all you can of our Lord's Word. Ashamed to ask is afraid to learn, this is not ignorance, but stupidity.

Call me what you wish, what you call me alters me not. We are not judged by another's attitude.

As our Lord is our Father and we his children, when we go wrong he chastises us. This is not for his pleasure, rather for our profit. All things our Lord does and allows, is for our profit.

Our political governmental school system is of satan. Children receive ignorance through "education."

There are many truths Science has ridiculed and rejected. One such is the creation of life. If man had evolved there would have been many, many missing links. Another scientific lie is the elliptical orbits of planets. Planets cannot have closed orbits. The big bang is another great lie. Even another is gravity. Gravity energy does not originate from mass such as planets, stars, galaxies, etc. Mass absorbs a small part of the energy of the universe to effect gravity. Some of the most educated people cannot understand the electromagnetic energy of light being both a wave and a particle (read the Universe of Energy), but I can do that, and I'm a dummy. And, if the most highly educated (trained) on earth cannot understand the simple physics of the universe, then they are far from faith, and even farther from understanding God's Heaven.

The Spirits of our Lord are invisible, but he manifests himself in the earth, the sun, the stars, and all living souls and creatures. He paints himself as the sunrise and sunset, the leaves in autumn, the blue sky, the nourishing rain, the crops of wheat, and the Spirit of love and faith. Look at each other and you will see God.

Ignorance of God's Will pierces a heart like a sword, dividing the spirit and the soul. Giving not our Father pleasure is being an antichrist and subject of satan. To know not God's Word is to jump into satan's bed. To do not the will of our Lord is to serve the prince of sin.

To be still, or to be quickened, is our choice.

All things we have are not ours, but is given to us by God.

Be not jealous of the rich, for they are tempted greater.

Science is a window to God's universe, but not his Wisdom.

As for God's Word, it is more profitable to learn one verse than to "read over" many chapters. It is better to understand just one verse than to misunderstand the whole Bible.

The spirit world and the flesh world are not truly different dimensions. The Lord is my Father, and I fear not what my brothers may do to me.

Be suspicious of those who claim the truth. No one knows the truth. One only knows a small part of the truth, at most. The truth is all knowledge (wisdom). No one knows all wisdom, save our Lord. All wisdom comes from God. We know little truth, and much lie. Anything outside the truth is a lie; therefore, half the truth is a whole lie. We are not wise.

To Judges, Lawyers, and politicians: If you are not concerned with my laws, I am not concerned with your laws. My laws are from our Lord God, your laws are from your lord satan. As I violate your laws you may, at most, kill my body, but as you violate my laws, your body, soul, and spirit may perish in the lake of fire.

Would you like to have a good life for 100 years? Would you like to have a good life for 1000 years? Would you like to have a good life for 100 years, and then a terrible life for 1000 years? Or would you rather have a terrible life for 100 years and then an ecstatic life for 1000 years? The choice is yours.

If the most important thing in your life is your life and not God, then you will lose both. If the most important thing in the universe to you is God, then your life shall be forever, and forever ecstatic. Why do we spend our precious time on sports, entertainment, play, and work, rather than on our Loving Lord? What do we really want most, eternal ecstasy or a few good years on this pitiful earth? Nothing is as important as eternal life, yet we toss it away like used chewing gum. We are stupid indeed.

There are no strangers, yet we know not each other. We are self centered, egotistical, supercilious, greedy, and full of lust. We are of God, yet some follow satan, clinging to sin. Our puffed egos destroy our shrinking souls. Only God, our Creator, through his mercy, can save us. Even then he allows us to choose our own destruction, although he will give us ways to repent and to find our way to him.

Those who obey God's Law still sin. Not so much for what is done, but rather for what is not done. Remember all the chances you may have had to help others, or give a needy gift, or even failed to pray for someone who needed it very much. We can not carry the world, or cure everyone, but we can contribute our works to those who need love, who need God, who need Christ, and saving of their souls. If just one Christian each converted a heathen to God there would be great joy in Heaven, and twice as many holy brothers and sisters as we have now.

Reverse discrimination is rapidly growing, partly due to the Anti-Defamation League. There should be no discrimination, forward or reverse. Persecutions of Christians in the world are now nearing the top of the abuse and discrimination list.

Olive

Blessed be the Olive tree, for it is Holy indeed.

Olive pronounced in Greek ("elaia") means "God of Jehovah" in Hebrew.

Olive pronounced in Hebrew ("zayith") means "God the Stone" in Greek.

Olive pronounced in English ("ahlifth") means "anoint" in Greek.

Olive pronounced in English ("ahyith") means "to teach," or "govern," or "bring forth thousands," or "friend", or "gentle," or "guide" in Hebrew.

The first part of Olive pronounced in English, "O" ("ah") means "Jehovah" in Hebrew.

The second part of Olive pronounced in English, "live" ("lifth") means the "Stone" in Greek.

Can spirits kill spirits? God is a spirit (seven spirits in one), and he will destroy the spirit satan.

There is nothing more dangerous to our soul than our own lust and greed.

I cannot understand why many, many people are so afraid to die, yet they toss away their eternal life. When you have eternal life you fear nothing that man can do to you.

The entire Bible is written around Jesus, the Christ and Messiah. Even the Old Testament refers to Christ Jesus, directly and indirectly. If you do not realize this then you will never understand the effect, and affect, of God's Word to us, nor will you ever achieve the wisdom contained within it. Wisdom comes only from God, and his Word in a special letter of instructions to us. Great works in a book are only blank pages if unstudied. From Genesis to the end of Revelation the histories in the Bible construct an unending true story.

People will do most anything to save their short lived life, but do nothing to save their eternal life.

"We have been told of earthly things and do not believe, how can we ever believe if we are told of Heavenly things?" (adapted from John 3:12)

To understand the Word of God you first must realize the "Light" in John 3:19 is the same as the "Light" in Genesis 1:3. Both are Christ.

What people think isn't what is. We are all deceived.

Your thoughts are not what you think.

If you wish to look smart, surround yourself with dummies.

If you wish to be smart, surround yourself with intellects.

If you wish to look big, surround yourself with midgets.

If you wish to be big, surround yourself with food.

If you wish to look handsome, surround yourself with ugly.

If you wish to be handsome, surround yourself with truth.

If you wish to look good, surround yourself with helpers.

If you wish to be good, surround yourself with Christ.

If you do not imitate Jesus you will follow satan.

If you do not follow Jesus you will imitate satan.

If you do not adore Jesus you will worship satan.

If you do not worship Jesus you will adore satan.

If you do not dedicate your life to God, and His Christ Jesus, you will dedicate it to satan and his devils.

The chaff is blown away by the wind, infecting others.

However you go in Christ's way you will be judged by.

Whatever you do to others, you will be judged for.

No one needs to know, or suspect that you are God's child. Do not create a credibility gap. Attend church as others, tithe to the church you know teaches the true Word and not man's fables and traditions. Celebrate traditions conservatively, not willing to worship anything or anyone other than God and Christ Jesus. Teach your children the true Word, they will learn man's word from their peers. There are many misinterpreted words in the English Bible. Search the scriptures and learn the true Word. Help others, especially the poor and the afflicted, in private if possible, but help with all and any means you can. Give to the poor, cheer the ill, save the unrighteous, avoid the darkness and those who desire to control (control is witchcraft).

Whenever you do not do as Christ did you will be judged by him.

Wherever you plant the seeds of Christ you will be judged by him.

Whoever you treat as Christ did you will be judged by him.

Civil laws are not of God's laws. A few laws may coincide with gospel law, but most oppose it.

A "man" of Jesus Christ, our Savior, is not of weakness, but of God. He fears nothing of this earth, or satan. Death to him is life forever in ecstasy with our Lord. He suffers persecution, pain, and poverty. These are blessings of our Creator. It is the rich that blaspheme our Lord, bring the poor to the judges, take money from the poor and save to the rich. The rich spend money on the rich. They deceive the poor, to control their minds and bodies. The rich are of familiar spirits, devils, and darkness. The poor may not be righteous, but they are not in league with satan, performing witchcraft, as are the rich. This will soon change.

All the money of all the rich cannot buy eternal life. The only way to buy this is with your temporary flesh, given to God, in the way of Christ Jesus.

Opinions of Americans are not their own, they are brainwashed. If you know and admit you're brainwashed, and try to correct it, that is wisdom. But, if you do not want to learn the truth, that is stupidity, and stupidity is fatal.

Those who believe that they cannot be brainwashed are perfect subjects.

Those who believe that they cannot be deceived are perfect victims.

Those who believe that they cannot be enslaved are perfect slaves.

Those who believe that they cannot be damned are perfect fools.

Can spirits kill spirits? Yes! God is a Spirit, actually, Seven Spirits. he can destroy your soul, cause you to perish forever. However, satan cannot harm you, if you are a child of God. Should you have the mark of our Lord in your forehead, satan cannot touch you. He may try to whisper naughty words in your mind, but he cannot even read your mind as our Creator does. Yet, satan has a very high "IQ." An intelligent person may have an IQ of 130, but satan has an "IQ" of 130 billion plus. He knows your wants, your moves, your signals. If a psychiatrist with an IQ of 130 can foretell your personality, then imagine what satan can know of you. He doesn't need to read your mind to know what you will respond to, especially if you are not one of God's children.

Talking, Praying:

Sometimes they are the same, sometimes they are not.

As you pray to God you are talking to him.

As you talk to God you pray to him.

As you talk to the saints (or angels) you are only talking.

As you pray to the saints or angels you are still just talking.

Only to God and His Christ shall we worship, and to nothing, nor anyone, or anything else.

Our Creator made:

the dark, the light,

the wrong, the right,

the down, the up,

the small, the large,

the sad, the happy,

the bad, the good,

the evil, the righteous.

All things have an opposite, until after the day of the Lord. Then all things shall be good with no bad, no wrong, no small, no down, no evil, no sad.

Why do people pray to saints? Because they ask the saints to intercede for them. Who can hold fast the face of God, our Creator, better than the angels and saints? We are far below, and so much less holy than the saints, yet we ask one another to pray for each other. If you believe that a prayer to the saints or angels (saints are angels) is worship, then we also worship one another. When one kneels in front of a statue he/she does not pray to the statue, but rather to the saint the statue depicts. Yes, we can pray direct to our Father, and we should constantly, at least talk to him and thank him for his blessing and his mercy he has given us all. And, we as sinners need all the help we can get, especially from holy saints.

A cross or crucifixion is not worshiped. They are symbols and reminders of the Lord, our Creator, and his son Christ Jesus.

We, as children of God and subjects of Christ, should obey civil laws if they do not offend the teachings of Christ, or reduce our Christian credibility. But, Christians should not comply with civil laws that are contrary to the Teachings of Christ, as you do, you wish "God's speed" to the evil. Pay not the fraudulent taxes that enslave our children and grandchildren, oppressing God's Word. Pay those taxes only that support Christ's teaching.

Fear not those who call themselves Jew, and are not, for they are of their father satan. Support not the products that cause income for the antichrists, and the meddling in sin. Shut not your eyes and ears to the ads of evil, learn of these, and list to avoid. Contribute not to the synagogue's of satan.

Beware the English version of the Bible, it contains many detours, winding paths, adverse to the old scriptures. Study the Bible and the scriptures with the old interpretations in Chaldean, Hebrew, and Greek.

Let not man lead you from our Lord and Christ Jesus. Know that none may obtain Heaven unless they have come down from Heaven. The fallen angels are no longer of Heaven, but now of earth. The night, the moon, the darkness, the sunset, and the month are of satan. The Light, the Day, the Year, are of Christ. Remember this as you read the Bible. As a day begins at evening it is unholy, morning is the beginning of everlasting life. A new life offered for us all through Christ Jesus and his sacrifice of himself, who was perfect without sin, who carried all our sins on his cross.

Christ changed the symbol of torture and death into a symbol of eternal love and life. The cross and crucifixion are reminders for Christians of this, and who we are to serve. They are reminders to love our brethren, our families, and our neighbors. Reminders to obey our Lord's new covenant, and to thank him for his mercy, and that he is in charge. And, at the end of the Book, satan and his parasites lose. And also, as a sign to others that the wearer is a Christian who loves God, and not ashamed of Jesus Christ, our Lord. And, saying to the brothers and sisters in Christ, I trust in the Lord, my God.

The ones who wear sports caps, logo shirts, and now, Mickey Mouse watches, along with those who are fans of entertainers and other "heroes" are worshiping idols and false gods. Those entangled in businesses, stocks, and gains over others' losses are witnessing for satan. Only fair trade is conducted by true Christians.

And, those who want control of others are in league with satan. Even as minor as to place small burdens on a spouse (physical or mental) to influence him or her for control, is against our Lord's face, because this is witchcraft. True Christians tell not others what they should wear, or do, or love, but only advise. This does not mean spare the rod for our children, they need guidance, leadership, and discipline. Raising offspring is not witchcraft.

As we wish to please our Lord, we must plant his seed in those that it may sprout. Beware of new illusions and "near truth" lies about Christians and the Bible. No, my brother and sister Christians, crosses and crucifixes are not idols, only memories.

Even God, our Lord and Creator has given us instructions to build statues and reminders, even in his own words. A few examples are:

Numbers 21: 8-9

1 Kings 6:23, 27a, 29, 31-32

1 Kings 9:3

John 3:14

Colossians 1:15

and the most famous of all: Exodus 25:17-19.

No, my brothers and sisters, prayer to the saints is not worship, but a cry for help.

Protestants ask why Catholics pray to saints. Catholics do not worship saints or statues, they ask them to help and to pray. Protestants ask others to pray for each other. Both Protestants and Catholics are unholy sinners, far from God. Why would unholy sinners ask other unholy sinners to pray for them? Would not the saints hold the face of God faster than unholy sinners?

If I was in hell, yet I could see my children and their children, and my family and friends in Heaven, I would have my own heaven, knowing their joy.

My worst hell is to know that I was the cause of someone being in hell, especially one of my loved ones. Yet I must carefully lead them to our Lord, save they turn away in spite.

When life turns sour, and things go against you, have faith, prayer, and trust. You may think our Lord has turned away from you, but he does not turn away, even though you turn away from him. We, as sinners, are egotistical, vain, greedy, and hateful. Even so, we love our children enough to die for them. God, being perfect, loves his children infinitely more. Faith, prayer, and trust create the most beautiful experiences in the world.

Faith is the most precious love in the universe.

Prayer is the most powerful action in the universe.

Trust is the most patient effort in the universe.

These give patience, humbleness, and greatness to the saints. These will make one tall and brave in stature, wise and knowing in spirit, and very loving in heart. The love for our Lord is the most powerful of all things in Creation, except for the love our Lord has for us.

When loved ones pass away, death is not who comes for them, but our Lord, the God of the Living. Remember your loved one will become a spirit, an angel of God, looking after you and your remaining loved ones.

Our God is just and will answer all prayers that come before his face. Always love our Lord, for he always loves you. And, in a moment or two, we will be with our loved ones, and the One who loves us all.

Focusing on the faults of others wastes the mind, and the soul.

Although we are dumb, we utter vain noises.

Are we not mud (water and dust)? Then with what pride do we deceive ourselves? What glory have we but in the love of our God and his Son. Even Jesus, the perfect man without sin, glorified only our Father, and not himself. Jesus only asked the Father to glorify himself so he could glorify the Father.

To forgive someone you probably must have judged them. We have not the right, nor the grace to judge anyone. We must forgive before we judge.

Death does not come to true Christians, but our Lord, the God of Life, comes unto them.

Revelation is not a concern of the world, but of each individual soul, as they will be in Heaven, hell, or elsewhere. Each person must encounter their own tribulation.

One microwrong erases a million megarights.

Maintaining the grace of God and Christ is like cutting down bad weeds; you must continuously chop at evil or you will be overgrown by it.

Oppressing evil is like swimming against the current; you must keep stroking hard to maintain ground, but if the current is much too strong for you our Father and merciful God, will push you to shore.

Even an atheist, who believes not in our Lord, or his Son, yet walks in Christ's way of love, has a greater chance to enter Heaven than a "Christian" who does not do Christian deeds.

How can we, being unholy sinners, correct anyone who reproaches us, as we continue to offend our Lord.

Always have patience, and do not consider who gives you trial, because trials come from both the good and bad, the holy and the evil.

Eternity can be won simply by victory over yourself.

Think in spiritual ways, not flesh. To lose someone you love is sadness, but only for yourself, this is selfishness and vanity. If you lose one who is evil be sad for them. If you lose one that is holy be very happy for them.

One who manages to think in spiritual ways will soon have the Holy Spirit with them.

I would that I have nothing in Heaven than I own everything on earth. For if I owned all on earth I would have an agreement with satan; then, soon I would own nothing in hell.

God, our Father cannot be evil. He will not draw us to temptation, although he will test us with love and patience. Our ego is the main gulf for us to conquer for the reward of eternal blessing. We seem to believe that the things we accomplish, the master piece we write, acquiring of wealth and power are of our doing, our intelligence and by our own strength and knowledge. We are nothing without our Father. Remember, if our Lord turned his head, or forgot us for even the smallest part of a second we would cease to exist.

As God is eternal so is his intelligence, knowledge, strength, mercy, forgiveness, blessings, and love. Why do we often think we are worthy of these eternal gifts? Is it our ego, our selfishness, our greed, our laziness, our lusts? Yes, it is all these, and also our stupidity. We seem to care little of our origin, or the Holy One that created us. Remember, we are on trial in the flesh.

We have no fancy lawyer, or shyster to "rescue" us from the trials of life. We have a Savior that gave his life for our souls and spirits. We, ourselves must endure the life in the flesh if we are to enjoy the ecstasy of life in the spirit. For Christ Jesus showed us the way, his way. There is no other way, but satan's.

We must sing in the choir of our Lord, dance at the beat of our Lord, work in the works of our Lord, weather in the words of our Lord, live in the life of our Lord, and walk in the steps of our Lord. To follow him step by step, word by word is the way.

Are we in warm water? If you know anything about cooking, especially cooking frogs, you will understand that if you throw a live frog in hot water he will dash and splash and jump out. But, if you place him in cool or luke warm water he will stay, even as you slowly turn the temperature up. This old frog will have a stupor come over him, and he will fall asleep and wake when it's too late, for then his goose is cooked. We are frogs and satan is the cook. Yes, folks we are in hot water. This applies not only to our faith in our God, but our trust in government.

The difference is that our God has not deceived us, but our fellow humans have, because God is not affected by satan, just us.

The proof of the frog pudding is the treatment of our young and our old. Our society approves of destroying the unborn, and the elderly sick. We spend billions of dollars on sports, entertainment, and cosmetics, but little to none on dying children and elderly. Our so called "government" spends illegal billions on treatment for AIDS, but nothing on gout or arthritis.

All this is leading to the one world beast, as told in the scriptures of old and new. The scriptures are on tract. Why do we not learn them? Do we not want the most precious gift ever offered?

There are three dimensions, Heaven, Earth, and Hell. Heaven and Hell do not touch. Earth reaches the bottom of Heaven and the top of Hell. Earth is in between Heaven and Hell. It is a part of each. On earth there are opposites. There are good and bad, high and low, front and back, up and down, right and wrong, etc. In hell there is only bad. In Heaven there is only good. Somewhere our universe merges with each, being the Light and the dark. Thus, all universal things endure opposites. Yes we are in between Heaven and hell. In which direction will you walk?

If you listen to fools, you become a bigger fool than they.
Remember, the Word of our Lord is only foolish to fools.
Why do we worry about what others do, we will not be judged by their sins.
When we are with our Lord, our Lord is with us; who then can come against us?
As you sin against our Lord you violate your own soul.

Science has gone astray, as according to the ways of satan. Scientists, on super ego trips, claim science prevails over the Word of God. Firstly, the Word of our Lord existed long, long before science. Secondly, science is a function of the way of satan, to lead astray. Thirdly, evolution is not based on the scientific method; (1) observation, problem?, (2) goal, (3) search and explore for evidence, (4) generate alternative solutions, (5) evaluate the evidence, (6) make an educated guess (hypothesis), (7) challenge the hypothesis, (8) reach a conclusion, (9) suspend judgment, (10) take action, (11) induce motivation, methods and processes of skills, (12) also always challenge results. There is no evidence for evolution that is not contrived through evil, misunderstanding, fraud, and greed. It is the same for Cosmology.

An open mind contains an infinite amount of wisdom.

I have much ego, too much; but, I cannot approach the eyes of scientists, politicians, or celebrities.

Look around you, you can hear, see, smell, taste, and touch our Lord, for all things are of him and he in them.

We are all in one Christ, and our Lord, for we are his children. Remember, God loves his children, but he does not love some of the things we do.

Beware the religions of man, for he may do many evils under the guise of holy names. There are many truths in the Bible that are taught as untruths, and many lies of man taught as gospel.

You should not own your Bible, your Bible should own you.

The love from our Lord even exceeds the most powerful thing in the universe, prayer. You may remember that they are both larger than the universe, they are of Creation.

You must rid yourself of the evil thoughts in your mind before you can rid yourself of the evil spirits in your thoughts.

Faith or obedience? Which does our Lord want from us? First, we will not have obedience if we do not have faith. Second, we will not find true faith if we do not have obedience. Obedience is the keeping of our Lord's Word, as which we are judged on. If we do not keep our Lord's word we sin. Our sins are what we are judged on. Our sins consist of what we do, what we say, what we think, and what we do not do. Our actions and works are of these four things. Our sins are of these four things. And, our judgment is of these four things. These four things are made of faith and obedience. Of faith or obedience we must start with obedience in order to obtain true faith. And in true faith true obedience will evolve. This is the only true evolution in existence. Faith or obedience, for all practical and holy purposes, are one and the same.

God tries to teach us that we should walk in His Word, for anything that cost nothing is worth nothing. No trials, no triumph. Our Lord's trials always have rewards for those who triumph.

Beware, things for free are often too costly.

Trusting our Lord and tempting him are often confused. Letting God lead your life in His Word is trust. Letting God save you from foolishness is tempting him. Catching you if you fall is truth, but catching you if you jump is false. Praying that God will catch you if you fall is trust. Praying that God will catch you if you jump is tempting him.

Millions of people spend billions of dollars on psychiatry analysis help. We have the greatest Psychiatrist of all, our Lord. If we would spend an hour, or just a few minutes a day attending a session with our Lord we would have little, if none, psychological dilemmas. The price is right. Our problems are solved by our Creator, that will only cost you the rest of your life, which will be eternal, and in ecstasy.

The UN, the Israelis, the United States, Inc., Britain, Clinton, Bush, Congress, the brainwashed Americans, oil companies, and the World Order are all against

Saddam, Iran, Syria, etc. Does this not tell you something? Are you part of this group?

Grandparents are like teenagers, they know everything. The slight difference is that grandparents know everything through experience, and teenagers know everything through conceit.

Why do African-Americans call themselves black? Black and darkness depict the absence of Light. Darkness is the side of satan, where there is no Light. As you walk in the dark you cast no shadow. But to walk in the Light is to cast a shadow. Christ Jesus is the Light, therefore we should always cast a shadow. If we do not walk in the Light we cannot cast a shadow, and when we cast no shadow we are in the dark. To cast a shadow on another is to plant the seed of righteous love, and to aid in saving a soul and spirit. If one does not cast a shadow he is not an example of a true Christian.

The vanity of glory is pride, a great sin. This sin creates many other sins, violating the Love of God.

A statement that is not all correct may be all incorrect. *Truth that is not all truth is all lie.* A small lie inserted in a large truth can create a total false statement. Most of the time a statement that is not all true is all lie, whether it is done with intention or by accident.

Do not argue with satan, he will turn you inside out. Even the Archangels do not argue with him. The only remark to make is "God rebukes you."

"Bad" kids often become corporate brats (WO, politicians, lawyers, etc.). Our schools, public and private, grade and college, train and encourage our children to become adult brats. This is by design.

Laws:	made by God
Regulations:	made by congress
Codes:	made by agencies
Statutes:	made by states
Ordinances:	made by counties and cities
Honest contracts and agreements:	made by free people

If you believe that there is no God, and you give your trust unto man, then you shall compare souls. Your plow will become a sword, your loves will desert you, your existence will be short.

Fear not the pains of persecution, for to be "crucified" is to glorify God.

People do not wish to know the truth, because it will cause a change, and people hate changes. Worse yet, it may cause one to think.

How can we judge others when we cannot weigh our own ways?

The Ways:
    recognition of sin
    repentance
    forgiveness
    the Word
    Love for our Lord
    Love from our Lord
    Grace and blessings from our Lord

Why is there so much suffering in this world? The answer lies in who is the prince of this world: satan (lucifer, the beast, the big horn, death).

To take a chance. We all gamble with many things, especially in decisions, our lives, and our souls. We greatly improve all our "odds" as we love God, and follow Christ. But our odds go to zero when we turn away from our Father and our Savior. Why gamble with about 80 good years of this miserable world against an eternity of ecstasy? We can stack the odds for or against ourselves, it's our own choice. If you want a sure thing, go with God, he controls the odds.

If you knew that a specific horse was going to win a race, would you wait until the very last minute to bet on it? Maybe the window would be closed before you reach it. Then, why do you chance the loss of the most precious thing in Creation, forever in the arms of our loving God? We do not know when our Father will require our soul. Turn to God now, before you perish. Do not take a chance and live on the edge. The edge slices and cuts you away from God. No matter which side of the edge you fall, satan is there to stand you on another edge.

Again, the love for our Lord is not a religion, but a reality.

Don't chit chat, leave babble to the Babylonians.

We direct matter of no value to urgency, but slight the way to God.

We know little of the past, and nothing of the future, what knowledge can we judge with?

As Jesus went among the wicked to save souls, satan must go into the blessed to destroy souls.

Remember, the Bible is a book found in the non-fiction section.

It seems as though people are anxious to please their masters, bosses, supervisors, etc., but they allow little time or effort to please our Lord God, the one who truly matters, in all things. No man can give another anything that doesn't perish. No man can offer a single soul, or spirit eternal life, or even eternal death. We can only earn that our selves as a gift from God.

Will you love our Lord so he may breath upon you as kindle and give you the fire of life? Or shall you remain in sin and ignorant of his Word, as to anger him that he may breath upon you and blow you away as leaves soaked in sin?

God is a Consuming Fire, the Light of every burning life. He can consume you as eternal life, or as ashes, it is your choice as he has granted you. You must either love the Creator or love the Creation. We have but little time to decide, don't get caught with your love down.

To read the whole Bible will surely improve your way, but to read, study and understand just one verse will greatly improve your whole life.

Be proud of nothing, nothing at all. Whatever you have is given to you by God, and it is also temporary. Pride causes many, many sins, including greed. God does not give you pride, this is whispered in your ear by the children of satan. If you can overcome pride you can beat satan. To beat satan is to earn eternity with our Lord.

The word "I" is the most used word in any language.

Money is the mark of the beast, not a chip, or stamp, or number, or ID, but just plain money, money in your hand, and in your thoughts.

There are many, many self interpretations and invented meanings of the Bible. Most are wrong and self centered. However, every chapter of the Bible has at least four meanings (maybe even seven).

The first is the Apparent meaning. This is revealed by interpreted words of scriptures. This meaning may lead one to God, although there are traps and deceptions in the interpretations.

The second is the Illusion. One believes they have found the true meaning of a phrase, verse, or even a chapter, then they even have their belief confirmed by others. The greater the number that confirm this belief the greater the illusion.

The third is the Inference. This is found only by searching scriptures thoroughly with painful, tedious work, and learn the colloquial language of thousands of years ago. Once in a while a secret is discovered. Logic, deduction and reasoning is a major factor in determining the truth.

The fourth is the Concealed and obscure (secrets). Only the Holy Spirit will afford this truth to the chosen elect. These secrets cannot be attained by flesh alone. The secret of these in Hebrew spell Garden. There are three more meanings in scripture that we will not know until Heaven reveals them.

The problem with us is us, not God. Blame nothing on our Lord, he is perfect, we are not. It's the creations that are prideful and greedy.

We are always searching, but we will never find what we are searching for because we do not know what we truly want. But, when you find Christ Jesus you have found the only thing you need in life, and especially in death. Although, when we find Christ we continue to search for more of him.

Do those in Heaven want to look back to earth to see what's happening? Why would they? If you sold your house and moved from an area you hated, then bought a new house you loved in a neighborhood you enjoyed immensely, would you travel many, many miles across country just to see your old house and area?

Wisdom is a fleeting gift earned only through the love of God.

God allows us only what we can handle, good and bad.

Again, remember if you want to see God look at each other.

Do not look at any person as a person, look at a person as a soul.

Again, fear not the God that loves you, fear his wrath.

Christians cannot understand their God without the Old Testament. Jews cannot complete their laws and find salvation without the New Testament.

To discuss religion is profitable. To argue religion is foolish.

To argue with yourself, you will not lose, but then you also will not win.

If you have not learned from an argument you have not won.

Those who believe they are not sinners are sinful in thought, indeed that is sin itself. Who can glorify God all the time? If your faith and works do not glorify God then they are in sin, and only temporary.

Works without faith are false works. Works from faith are true to God. When you have true faith your works are true. If your faith is slack, your works are under color.

The Lord's people are not cowards, they fear nothing of man.

If I am at the top or bottom of the poll, God and I together can beat any odds.

Do not pray to be saved from the ones who would change your way, but to rebuke the ones who would turn us away from His way.

The Bible was written through 1500 years, and rewritten another 2000 years. The Bible has been burned, misinterpreted, mis-interpolated, always misread, mistranslated, separated, divided, and shunned, along with all the terrible things that can happen to a writing happened to the Bible. The Bible has been destroyed in as many ways as possible, yet it's essence still is, and will always be in existence. Its mystery of attraction and absorption grows every day.

Each and every person has a different opinion of each and every chapter, paragraph, sentence, and word, yet the Bible's overall meaning remains.

No one knows the Word of God (Rev. 19:12, 13), yet we attempt to convince others of our opinions according to our beliefs, right or wrong, or in between. ***No one*** knows the Word of God.

What profit have you if you own the world for 100 years, but lose your soul forever?

God gives us the seed to plant, the water to feed the seed. God creates the seed, the water, and the increase.

We know of the past, and only a guess at the future. When Christ Jesus comes, the past will be discarded.

Evil friends create evil friends.

Do not listen to any word that does not honor God, for what does not glorify God is sin.

Christ Jesus's last words:

"Father, forgive them for they know not what they do."

"Verily, I say unto thee, today shalt thou be with me in paradise."

"Woman, behold thy son."

"Behold thy mother."

"I thirst."

"My God, My God, why hast thou forsaken me?"

"It is finished."

"Father, into Thy Hands I commend my Spirit."

If you have suffered without succeeding, someone else has succeeded. If you have succeeded without suffering, someone else has suffered.

When Christ was crucified God allowed the evil one to slay God's Lamb, rather than man to slay man's lamb. By death crucifying the Light, death was defeated. In other words: satan defeated himself when he crucified Christ Jesus. The Light was perfect always. Christ was the Beginning and the End, the First and the Last, the Light and the real Morning Sun. Christ could do what no other flesh could do, be born not in sin, and perfect forever. Christ is the Light in Genesis 1, and the Elijah in Malachi 4:5,6, and Jesus at the end of Revelation (Rev. 22:20).

In Hebrew Aleph is the first letter (means Father), Tav is the last. In Greek Alpha is the beginning, and Omega is the end. Christ was in the beginning and will come in the end.

If you are totally in Christ, you are totally free. If you are partially in Christ you are far from free. When you are enslaved by Christ you are free. When you are free by satan you are dead. When you are in Christ Jesus you need not the Law.

Money is also in denominations, but the only denominations we should seek are those that seek God and His Son Christ Jesus.

Thoughts are not what you think they are.

Works without faith is still works, but only works. Faith without works is false faith.

Truth is what you believe, right or wrong. Ignorance is no longer bliss. Only the knowledge of God and his Word is bliss.

When you hurt another you hurt God, because we are all part of God.

Who prevents mankind from attaining everlasting life? satan and his minions!

Who is satan? lucifer, a cherub!

What was set at east of Eden (gate) to prevent mankind from putting forth his hand and also to take the Tree of Life, and eat, and live forever (Gen. 3:22)? A cherub and a flaming, turning sword (a seraphim)!

Gen. 2 ... And the man (ha-Adam) knew his wife (Eve). And she conceived. And she continued to bear his brother, Abel.

Note; Cain and Abel were half brothers (conceived of different sperm), Cain of lucifer, Abel of Adam. Adam was not the first human on earth, he was the **first in line to Christ Jesus**.

Seth was conceived to replace Abel as the second in line to Christ Jesus. Christ Jesus would allow all mankind to be forgiven of sin, all who would put forth their hand and take from the Tree of Life.

Christ Jesus is the Tree of Life, satan is the tree of knowledge. We still have a choice.

Thus, Moses was writing of Jesus as the light in Gen. 1:3-5, and satan as the serpent and fruit in Gen. 3: 1-6.

The human mind cannot conceive infinity, much less God's other mysteries.

Scientists claim they know the secrets of the atom and the universe, yet they know nothing of their souls. Our souls will survive by love, not knowledge.

The study of knowledge should be the knowledge to overcome our sins and passions of earth, flesh and material.

It is easier to avoid sin than escape death.

It is much harder to define love than to feel it.

As your mind shall live so will your soul.

It is very hard to love one whom you suffer for, but to suffer for one you love is effortless.

Our Lord knows when to give us trial and when to give us comfort, but he will always give us love.

The love of Christ supports all the commandments.

O' lady of lust, how far and clear I could see if I were not looking at thee.

When you boast, boast of the Grace of God, for it is he that has given you all you have and all you will have.

What you call another, others will call you.

Anything above humbleness injures your soul.

What chance have we to become worthy of God. Solomon was seven years building a temple, Moses made an Arc covered with gold for the Tablets of Law, Noah worked 100 years to build his ark. Many saints suffered much and died painful deaths for God. How little we give to God. We're sunk!

What are we thinking when we do not bend to help another? Are we too good? God is good and he stoops to help us constantly.

When you wish to suffer anxiety, anger, frustration, disgust, listen to idle chat, or the news of the world.

Where shall we find peace and refuge? Not in the world of vapors.

We cannot overcome nature, so we must overcome our passion of nature.

To accept advise is much safer than to give it.

For whatever you have done great in this world today, it will have been in vain by tomorrow.

Today you are here, tomorrow you are gone, and in a few years no one knows or remembers what you accomplished, even less who you were. All earthly treasures are split among the rich and have no meaning where you will be.

Vanity was given to us by satan, as vanity flourishes so does satan.

As the stupid beasts we are, we bring death to our souls for the trivial enjoyments of our corruptible short life.

Anxiety is also in vain, we anticipate things that do not happen.

While we attend to high numbers, as our souls become zeroes.

Why do we go for the gusto which is whisked away in a moment?

We expect perfection in others, but allow foolishness in ourselves.

Strive to overcome the things in yourself that you despise in others.

When we trouble others we trouble ourselves tenfold.

Many times we seek answers and comfort from worldly things. Only when they fail us, do we turn to Christ.

Why confess your sins today when tomorrow you will repeat them?

If you are lucky enough to know your last hour, you shall grieve deeply of your wasted life.

If you do not attend the future of your soul there is one who will, satan.

The past is forgotten, the future is unknown, the present is fleeting and is the only time to save your spirit's soul.

The one with the most toys surely dies, but the one with the least shall truly live.

How can we have faith in evil flesh and it's material, yet reject Christ who lived and died for us?

Can anything in this world of vapors give you life everlasting?

We have choices, comfort or pain, light or dark, Heaven or earth, eternity or time, it is our choice.

As long as our soul endures our frail and sinful flesh we will suffer the material troubles of this life. *Let our flesh suffer troubles, not our souls, soon our flesh will vanish.*

There are miracles every day. We do not realize they occur because they are common in nature.

We are the natural part of our supernatural God.

If those who praise you are not of Christ, then how can you be proud? Those who praise shall also scourge, and soon pass away.

Do not allow witty phrases of evil silver tongues to lure you from the grace of Christ.

The evil proceed with wicked judgment.

The Kingdom of God does not come by words, but the power of Christ. Listen to his words and be of those with ears.

The wicked outnumber the righteous, therefore wrong judgments prevail.

Politics is evil indeed; its words are bright and light, but its deeds are dark and heavy.

The domination of people is done of deceit through satan.

Money is the idol created by satan. Greed is the creation of flesh.

The evil ones have no shame, no respect, and no compassion.

We who are earning wages are placing them in a bottomless boat.

The flames of hell feed our sins, and our sins feed these flames.

A merry eve brings a hairy morn.

They who live in pleasure all their lives will die in pain all their death.

If you believe it is hard to endure the pains of life, which is short, consider the torture and torment of the lake of fire, which is forever.

We have a choice, we cannot enjoy the pleasures of life and death.

Some would wish to be a witch, so as to make magic. Remember, a witch must make a pack with satan. If there be a satan then there be a God. Guess who will win the last battle.

If God should give no mercy to us in his Judgment, we would all be tossed into the lake of eternal fire.

The suffering we endure now is naught compared to the suffering we may be earning.

No motive can justify evil.

It is so much easier to say "I love you" than to prove it.

As you deceive others you deceive yourself. An evil one deceives himself as if he were another.

Do not crave the praises of man, you can do a thousand things to please people, but yet one small wrong will cancel all the rights.

Become pure of heart and thought, for you cannot deceive Christ our Lord.

Even satan draws to you through spiritual weakness by lies, half truths, love of flesh and material, pleasures of the past, and fear of the future.

We are all deceived in politics, law, science, education, entertainment, sports, work, news, religion, business, and all earthly categories.

Almighty God is our Maker, Christ Jesus is our Savior, the Holy Spirit is our Teacher, the Angels are our Guides, and the Saints are our examples.

Love God, Christ, and the Holy Spirit so that your sins may be blotted out rather than your soul. This is the Holy Trinity and the Supporters we should die for to live with.

Evil dims the memory of peace and writes its own script in your mind.

The eternal Judge is not robed and dressed in silk, nor wishy washy of law, nor under color, but truly knows our guilt and innocence.

One of our greatest trials in patience is moving in traffic. We enter trial after trial. There would be a great advantage for our soul if we could master our patience while driving.

Maybe a child's death is beneficial to his soul. The longer we live the greater our sins.

To win a war you must always be ready to battle, your ears alert, your nose free, your eyes open, your mouth closed, and your heart with true love of Christ.

Fight hard at the beginning of temptation; the remedy may come too late. Victory cannot be had by flight alone.

Temptation is ever around, satan never sleeps, nor rests, as the flesh is subject to living trials. Christ will give to them who overcome the trials of life.

Christ never withdraws his love, but he will withhold comfort as we resist and battle temptation. As we win comfort begins again.

The more trials of life you battle the greater the rewards.

Christ our Lord, give us the wisdom to judge our motives and desires, and not to be deceived by satan or his servants.

Guard us and let everything be in accordance with your will, so we may serve you as you please, we are in your hands.

We are always enclosed in God's ecstasy, it is suppressed by satan's cloak. We arrive at true ecstasy only through the absence of satan.

Remember, we must prove ourselves in this life to be worthy of the next. Many trials await us all. To battle trials and temptations in life is to battle for the Grace of Jesus Christ our Lord.

Our Lord Jesus and the Holy Father do not injure us, nor tempt us. The suffering, trials, and temptations we endure are of the evil world, but is only allowed as God wills.

The rewards of Christ are far too great not to seek.

Have faith in a mountain, not a volcano. The mountain is steady and fast, taken many years to grow and change.

The volcano is like flesh, it will deceivingly befriend you, then destroy you.

Christ our Lord is the Way, the Light. There is little light in us, and what we may have we easily lose to the dark side.

If you truly desire to be with Christ, attend your soul and little else.

Christ grants graces, but we seldom return the favors. We cannot grant Christ grace, but we can give him gratitude and glory.

The grace of comfort is a great gift of Christ. Yet, without discomfort of suffering, comfort would be ever common place.

How can I turn my interest into your interest, my Lord, when all motives of flesh are selfishness?

As we share Christ's suffering we will share his comfort. If we die for Christ, we will live with Christ.

We shall find no other way to God, except through Christ and his cross.

We have a choice; To take up the suffering of the cross, or take up the suffering of purgatory. Which lasts the longest?

We suffer many hours a day, day after day, to earn worldly material and flesh, which easily vanishes. Shouldn't our suffering be for something lasting?

If all our suffering is for Christ, our reward would be infinitely greater than any comfort we find on earth.

The more one dies to self, the more one lives for Christ.

Your night will always be in peace if you spend your day with Christ.

As we have love of Christ, we may live in peace among passionate people.

As we love Christ the truth is revealed that all things on earth are worthless. The truth does indeed set us free.

Beware as we suffer for Christ we are not led by our vanity.

How vain is it to study the Trinity when we are vain?

Evil thrives on passion. The love of Christ is not passion, but wisdom.

We judge others by ourselves, yet we still put them down.

If we are patient Christ will heal our problems.

The rewards of God are too great to gamble for trivial material.

Christ said "If I send you trouble and affliction, do not be indignant or down hearted; for I can swiftly help you, and turn all your sorrow into joy."

As we travel in life we have many disguised enemies giving us attacks from all winds. Love our Lord or be severely wounded.

Always remember that there is no greater ecstasy than the touch of God.

Follow no one but Christ, for they who overcome us will bring us into their bondage.

If you master yourself, you are the master of the world.

Do not allow circumstances to control us. Maintain our inward freedom as our own master, not evil's slave.

When we have doubts in which direction to go, we should always seek the answer from one of the Trinity, and their Supporters.

To serve Christ, or to serve satan? That is the (stupid) question.

All the grace of the saints were given to them by Christ. he made the saints with glory and grace.

Oh Lord, if I were only worthy to be the dirt in your footprint.

We are beyond help by any creature in Heaven or earth if we are abandoned by Christ.

We may strive, but cannot attain virtue, because we are not angels or saints, and even they were not in constant virtue. But if we do not try we will degrade in grace continuously.

Christ is just and his judgments are right, and are to be respected, not discussed, because they are beyond comprehension by the human mind, as is simple infinity and eternity.

Why do we wish to be honored by man when all things they esteem are held as abomination by Christ.

We can give all we have, mind, love, soul, body, and this is naught compared to the service Christ has rendered to us.

What trials should we endure? The temporary trials of life, or the eternal trials of hell.

To pleasure our dirt body is to damage our spirit and soul. We are but holy souls wrapped in bags of sinful dirt.

How can we unclean sinners answer to any who reproach us, when we continuously offend Christ, and are deceived by satan constantly?

Christ our Lord:

> We are cold unless you warm us.
> We are hot unless you cool us.
> We are ignorant unless you teach us.
> We are defenseless unless you protect us.
> We are lost unless you guide us.
> We are proud unless you humble us.
> We are perished unless you save us.

To endure true patience we must not consider by whom we are tried. Our trials shall come by holy and by evil people.

We cannot win eternal peace without patiently enduring temporary suffering.

No pain, no gain. No work, no rest. No battle, no win. No love, no Christ. No Christ, no ecstasy.

To chose between the Creator or His Creation, is the true meaning of life.

Our hearts are resolved on Christ that the suffering forwarded by our enemies will be received with patience, and endured with pleasure. Therefore, we say strike at us satan, wound and torture us, for we yearn to suffer for our Lord Christ Jesus, as he did for us. We thank you for resolving our love for Christ. Tempt us as you will, for our love for our Lord only increases with each wound.

Our suffering on this Creation is but for a very short time. To give this little suffering for eternal comfort is a bargain indeed.

Our small bit of suffering on earth is truly not worthy of God's Kingdom.

Christ knows the troubles we face and fight, even in rush hour traffic.

Christ will give us consolation and peace in this life from time to time, a small touch of what is to come as we seek truth and battle evil.

To triumph over self is the perfect victory. As we control our flesh and are totally subject to Christ, we are masters of ourselves, and the world. This is true freedom.

Devotion and service to Christ during our times of ease is not close to the pleasing of Christ, as is patience and humility in time of adversity.

Christ's flesh is his Spirit, his blood is his Life.

Thank you for the small scourges I've received. Thank you for teaching me your Way, and to know the difference of time and eternity, flesh and spirit, death and life. Even though I deserve not your grace or your ear, you have listened and graced me with inner faith and comfort. Thank you for you my Lord.

If you don't devote a little time to God you will devote a lot of time to satan.

God does miracles, satan does magic.

The First in Line to Christ Jesus

	(Mathew 1:02)
	God
1	Adam (Eve)
2	Seth
3	Enos
4	Cainan
5	Maheleleel
6	Jared
7	Enoch
8	Methuselah
9	Lamech
10	Noah (Noe)
11	Shem (Sem)
12	Arphaxad

13		Cainan
14		Salah
15		Eber (Heber)
16		Peleg (Phalec)
17		Reu (Ragau)
18		Serug (Saruch)
19		Nahor
20		Terah
21		Abram (Abraham)
22		Isaac
23		Jacob (Israel)
24		Judah
25		Pharez
26		Estrom
27		Aram
28		Amminadab
29		Nahshon
30		Salmon
31		Boaz (Ruth)
32		Obed
33		Jesse
34		David

	Line to Joseph (Mathew 1:02)	Line to Mary (Luke 3:23)
35	Soloman	Nathan
36	Roboam	Mattatha
37	Abia	Menan
38	Asa	Melea
39	Josaphat	Eliakim
40	Joram	Jonan
41	Ozias	Joseph
42	Joatham	Juda
43	Achaz	Simeon
44	Ezekias	Levi
45	Manasses	Matthat
46	Amon	Jorim
47	Josoas	Eliezer
48	Jechonias	Jose
49	Salathiel	Er
50	Zorobabel	Elmodam

51	Abiud		Casam
52	Eliakim		Addi
53	Azor		Melchi
54	Sadoc		Neri
55	Achim		Salathiel
56	Eliud		Zorobable
57	Eleazar		Rhesa
58	Matthan		Joanna
59	Jacob		Juda
60	Joseph		Joseph
61	"Jesus"		Semei
		62	Matthias
		63	Maath
		64	Nagge
		65	Esli
		66	Naum
		67	Amos
		68	Mattathias
		69	Joseph
		70	Janna
		71	Melchi
		72	Levi
		73	Matthat
		74	Heli
		75	Mary
		76	Jesus
		77	Holy Spirit

Jesus said "I am the Way, the Truth, and the Life." (John 14:6)

Jesus said "My kingdom does not belong to this world." (John 18:36)

The Way, the Truth, the Life, and the Kingdom are not of this world,

therefore there is no truth on this world, because this is satan's world. When we choose this creation over the Creator we choose satan with his wrong way, lies, and death.

There is no truth on this earth.

Do you think the miracles done by Christ Jesus are now gone? But, you say Christ isn't gone. You are right, Christ is still here, and so are his miracles. All you have to do to see one is to ask Him.

Atheists are the true "jack asses." They are very smart rocks. Why do we judge others? Are we most intelligent, or most learned? If so, we still are not the most wise. We are not wise at all. If we rated wise from one to ten, we would be minus 12 (we, meaning humans). The only wise one on earth is Jesus.

Remember all "laws," other than the Ten Commandments, are of satan.

Why does the Catholic Church require Catholics to confess their sins? Do they not know their own teaching? They teach Jesus died on the cross to save us from our sins. Is there a "but" associated with the crucifixion? Did Jesus say he is saving us from our sins, "but" you must continually confess your sins to a sinner? Christ suffered much and died for us, to save our souls, not to save our confessions. God knows our hearts, so confessions to others are wasted.

In Chapter 1, Genesis, God gives mankind dominion over animals. We dominate the animal and plant kingdoms. On the reverse side, we cannot exist without animals, but animals can exist very well without mankind. We dominate them, therefore we need them.

Jesus called himself the "Son of Man" as opposed to the "son of satan." Cain and the Pharisees were the sons of satan. Adam was the first man in line to Jesus.

We can do nothing to cause us to go to Heaven. Christ Jesus has already done everything necessary for us to go to heaven, and he is still doing it. Therefore, we have nothing to be proud of, to boast of, to brag of as being blessed or righteous.

The nearest entity we can associate and perceive as God is gravity, since everything in the universe is created from gravity. The entire creation is God. The entire universe is God. But, the entire universe does not include creation, creation includes the universe, because creation is much greater than the universe. The universe does not include Heaven.

Pride, Greed, Envy, Gluttony, Lust, Anger, Sloth; these describes us well.

Sports and education are the greatest pushers of "Proud Of" stickers, which many, many Americans display on their cars. This indicates the massive amount of people that fall to the ruse of vanity, and pride.

Restaurants advertise "all you can eat" or "great amounts of food." This ruse pushes us to gluttony.

TV programs featuring multi-million dollar yachts and homes push us to wish and envy.

TV programs featuring sex push us to lust.

"Laws" and road rules, along with little or no responsibility due to insurance, pushes us to anger.

The price of goods and homes, along with excessive taxes, and advertisements push us to greed.

Government programs push us to being lazy and sloth.

All deadly sins are directly caused by, or backed by our "wonderful" government and politicians. Doesn't this tell us something? Who do our politicians follow, satan or Christ? They still keep telling us we are free. Oh, the lies we believe. A few of the major lies we believe are: Bible code, DaVince code, evolution, freedom, holocaust, global warming, ownership, rapture, and laws under color. Yep, we sure are gullible, yet we are proud.

If you love this earth you love satan. There is no truth on this earth.

Atheism is a magnified form of stupidism, and satanism. Atheists may not believe in God, but they follow satan. Evidently, faith is not always a good thing.

There are ones who claim they know that Mary, the mother of Jesus, was not a virgin. They claim Jesus was married and had children. They claim many denigrate lies of the Holy Family. These evil factions are also satan saturated, babbling fools.

People believe in science, astronomy, physics, etc. Most of these are at least fifty percent (50%) wrong and forty nine percent (49%) lie. This is far greater than the odds of the real God. Cosmology is almost 100% wrong and we just absorb these lies. God is infinitely right and we hardly listen. Our attention is occupied by satan, for we are entrapped in his world. God gave us free will, but satan is trying to control it.

You can buy happiness, eternal happiness, if you pay for it with faith, love, works, obedience, and righteousness.

Am I a Jew, or am I not? I worship the King of the Jews.

When you serve satan you profit in the flesh. When you serve God you profit in the spirit.

BOOK 4
POEMS
Silly Stuff

Mock Clock

Overslept clocks at relative velocity,
       Run not more slowly tick tock,
'Cept for those in animosity,
       Done of others a quick epoch.

Killing Time

Time slows not for our will,
       Never is it standing still,
Before us it won't reveal,
       Fleeting us, it will kill.

Link

Every occurrence anywhere
       Is connected to
Any occurrence everywhere

Spare Time

The entirety is with us, 'till it's cast
       Masking the future, marking the past
Of the wee quantum, to the infinite vast
       All time myths us, future and past
Each affects all, 'till the end ever last

Dun Did Dat

Go us 'er where and when
       Tho' none clever fair 'ave been
Lo, we ne're care go again

Vanity Of Life

Life is a short waltz,
          Looks are in vain,
Charm is but false,
          Love God to live again.

The Future

As men shall survive, supplied by the universe of power,
          To know whence comes the force that cleanses their giaour!
Their medicine, alas! fusion within their hearts core,
          To exceed the speed of light a million times a billion and more!
As wage of contention and joust in vain,
          Their physical construct precludes a wound and may never be slain!
The child's life will ne'er wane or ever conclude,
          Nor to be lessoned or fallen into desuetude!
As the light they so ultimately pass,
          Reveals the right of the future and the wrong of the past!

Flip Side

'Till man and woman join,
          Crave to become a single coin,
Still remain a head and tail,
          Save this sense to never fail.

Dream

A dream may be realized, and perhaps lost,
          Or merely be revised at half the cost.
If you dare brave this ammunition,
          You may fanfare a true premonition.
This image, although a blurred apparition,
          May focus into a starry-eyed vision.
Our mirage shows us much enlightened delusion,
          And awakens us with such heightened confusion.
A dream, surely, is never a waste,
          When dissolved, another takes its place.
Dreams have built the world, for good and bad,
          And ruled our feelings from happy to sad.

We all have dreams to wish we'd visualize,
      A cherished fantasy of the foolish and the wise.
Dreams of ours have forged machines,
      From conning towers to silver wings.
And, if your dream does not come true,
      Simply fall asleep and start anew.

For Ladies Only

What you hold before thee is not real, it's not what you see,
      Nor be what you feel.
Your senses may deceive what you hear, all you conceive,
      And a thought you most fear.
We're from Venus on a wild dream caper, no one can see us,
      As a filed ream of paper.
Although we are not dating, as you carry this page,
      We are now mating, in the very wisp stage.
Enjoy this ragin' love for a while,
      Wink at me page and shove me a smile.
I do thank you for the lusty hospitality, my name "Soo Anc Su,"
      You must be "Congeniality."
Whenever you immigrate by Venus, pause to go in,
      Endeavors to create life between us causes no sin.
Soon I must return through the few miles durst I came,
      Moon dust to learn of you, and our new child's first name.

Fluff Be We

As we look up into the sky,
      And see white fluffies floating by,
We are like them and they are like us,
      Both are created from heat, water, and dust,
Changing shapes into many known forms,
      To be blown about by violent storms,
Move silently or create a great noise and light,
      Gathering together, giving others a fright,
To great heights we often gain,
      Filled with fury and burst into rain,
Creating a vertical river and flooding the moor,
      Drenching the weary, the helpless, the poor,

Then, as we came we vanish into air that is thin,
        Others soon take our place, repeating our sin.

A Teen Miss-Advice

End your calls with "love you,"
        Even the ones you dislike too.
Always depend on friends or another,
        Never take advice from Dad nor Mother.
When you are stressed out,
        Break something of your brother's, no doubt.
Blame God that bad things be done,
        Take credit for a good day and fun.
Priority give to your friend,
        For Mom and Dad just more pretend.
Drive fast and tailgate,
        You will never hurt you or your date.
Be as lazy as you can,
        Never, ever lend a hand.
Problems you should never deal with,
        Treat them as only a myth.
Backstab and lie as much as you can,
        If caught just stick your fat head in the sand.
You can climb the latter to success,
        Using knives and lies with finesse.
If you can't think of anything smart, never,
        Always say "duh, whatever."
Use "always" and "never" when saying bad,
        About friends, Mom, or even Dad.
Never ever think ahead,
        Whine, pout and turnover in bed.
When you over talk on the cell,
        Whoa! Oh no! To say it farewell.
Forget math, science and school,
        You'll never grow up to be a fool.
Remember, you are clever and be greater,
        So, no one could ever see straighter.

Afterword

You have probably noticed that there is a common factor connecting Books 1, 2, and 3. The "Reality" is that we are being controlled through deceit in many theaters; such as politics, religion, education, science, economics, etc.

The primary, and essential, area used to control us "common people" is economics. The control of economics affords control of all other social arenas. The world order has this solidly, and completely, in their hands, and in our foreheads.

There is not much we can do to correct our plight, we are too greedy, and too reserved; except to write about these factions. Although, this presents a problem one must solve when writing about the circle that controls the world. You want to expose the lies and deceit enjoyed by these controlling factions; yet, for your own benefit you must leave them an escape route other than persecution or assassination.

Anonymity is their ally. Truth is their enemy.

FINI

Glossary of Terms

$_1\text{H}^1$ The symbol for the hydrogen atom nucleus. The "1" on the lower left side of the "H" designates the number of protons associated with this nucleus. The "1" on the upper right side of the "H" designates the total number of protons and neutrons combined to construct this hydrogen nucleus. This hydrogen nucleus has only one proton and no neutrons, but the hydrogen nucleus can have up to two neutrons and one proton.

absorption Energy Taking up, or in, energy (all or part of) by mass, and usually transforming this energy into another type energy (such as light to heat) or into mass.

acceleration Continuous change of velocity, either slowing down or speeding up, or a change in direction.

age of earth The estimated (calculated) time from now back to when the first two particles or protons came together to begin the earth.

amplitude velocity The maximum velocity associated with energy, either RMS or Peak values.

anomalies, gravity Variations in the Universe of Energy due to "shadowing" from stars, planets, and other particles, continuing novas, and many other phenomena occurring in the universe.

anomalies, orbit Deviations of planets or other particles from perfect or smooth orbits due to variations in the Universe of Energy energies.

arc sine (asin) Indicates an angle whose sine is …

Astro Dimension (AsDm) The universe and all its particles larger than an atom.

astro energy level The amount of energy surrounding earth and occupying all space and universes.

astros The universe, its particles and energies.

Atomic Dimension (AtDm) All the particles in the universe the size of an atom or smaller.

atmosphere friction The wind and its components turning in one direction (such as counter clockwise) causing adjacent winds to move in the opposite direction (such as clockwise).

atomic shells The appearance of a shell covering due to the velocities of the orbital electrons (see shell)

autumnal equinox The sun passing the celestial equator as its motion is from north to south, around September 21,22, or 23 (starts the beginning of autumn).

basic hydrogen atom An atom consisting of one proton orbited by one electron.

bell curve A statistical survey when plotted appears bell shaped, where the highest point in the center represents most numbered normal subjects, while the sloping sides represent the less normal subjects surveyed.

big bang A cuckoo cosmological theory stating that the origination of the universe began by a gigantic explosion, of which the resulting particles are still moving outward from this explosion. Sometimes referred to as the expanding universe.

binding energy The difference between the rest mass of an atomic nucleus and the total mass of each of the nucleon as separate masses.

build up, energy The cause of reflection due to energy entering a medium that slows it down while the following energy's velocity is still maintained in a faster medium. The reflected energy can be measured as a reflected coefficient.

bundles, energy See photon.

c Velocity of (visible) light energy, ~ $2.99792458E^{10}$ cm/sec or $6.7061663E^8$ mph.

c+ Any energy velocity greater than the velocity of light energy.

c- Any energy velocity less than the velocity of light energy.

calorie equivalent The conversion of a measurement of energy in one type unit as compared to the same energy as measured in calories.

collapse, wave packet The disappearance of a particle and its associated electromagnetic propagation.

compression The increased pressure on the leading side of an object moving through the UOE, as compared to the trailing side (see expansion).

concentration of energy The center of an object where the energies converge.

construct The physical and energy make up of a particle, or other objects.

continents The greater (or principal) land masses now protruding from (extending above) the ocean surface (Africa, Antarctica, Asia, Australia, Europe, North America, South America).

continuous fluctuations See anomalies, gravity

creation probability constant A standard average rate per cubic centimeter of the possibility of the origination of a particle in the universe (atoms created per second per cubic centimeter).

decay The transformation of one radioactive atom (nuclide) into another (which may or may not be radioactive) by emitting or absorbing one or more particles or photons.

dimension, astro See Astro Dimension.

dimension, atomic See Atomic Dimension.

density, electron The concentration of energy/mass of an electron measured in CGS system by gram per cubic centimeter (g/cm3). Approximately 9.72E^9 g/cm^3.

density, energy The amount of energy in a cubic centimeter, or measured as the amount of energy passing through a square centimeter in one second.

density, neutron The concentration of energy/mass of a neutron measured in CGS system by gram per cubic centimeter (g/cm^3). Approximately 9.72E^9 g/cm^3.

density, nucleus The concentration of energy/mass of an atom's nucleus measured in CGS system by gram per cubic centimeter (g/cm^3).
The formula being: dnuc = 9.72E^9/A^2. where: A is the total weight (mass) of the nucleus.

density, photon The concentration of energy/mass of a photon measured in CGS system by gram per cubic centimeter. The formula is: ph$_{dn}$ = 2.2415E^{-45}/λ^3.

density, proton The concentration of energy/mass of a proton measured in CGS system by gram per cubic centimeter (approximately 9.72E^9 g/cm^3).

e- Symbol for an electron with a negative charge.

e+ Symbol for an electron with a positive charge (positron).

E = mc^2 The formula relating mass, energy and velocity. In the CGS system E is in ergs, m is in grams, and c is in centimeters per second. (kinetic energy is: E = 1/2 mc^2). E = mc^2 is algebraically derived from the two formulae "E = hf and λc = h/ρ."
Where:
h is Planck's constant (6.6260755E^{-27} erg sec)
f is frequency in Hertz
λ is wavelength in centimeter
c is the average velocity of visible light (2.99792458E^{10} cm/sec)
ρ is momentum (mass x velocity)

ea3 Earth The energy absorption of earth in ergs per gram per second (approximately 1.133E^{12} erg/sec).

ea3 Earth moon The energy absorption of earth 's moon in ergs per gram per second (approximately 1.134E^{12} erg/sec).

ea3 Jupiter The energy absorption of Jupiter in ergs per gram per second (approximately 1.134E^{12} erg/sec).

ea3 Mars The energy absorption of Mars in ergs per gram per second (approximately 1.137E^{12} erg/sec).

ea3 Mercury The energy absorption of Mercury in ergs per gram per second (approximately 1.134E^{12} erg/sec).

ea3 Neptune The energy absorption of Neptune in ergs per gram per second (approximately $1.134E^{12}$ erg/sec).

ea3 Pluto The energy absorption of Pluto in ergs per gram per second (approximately $1.134E^{12}$ erg/sec).

ea3 Saturn The energy absorption of Saturn in ergs per gram per second (approximately $1.133E^{12}$ erg/sec).

ea3 Uranus The energy absorption of Uranus in ergs per gram per second (approximately $1.133E^{12}$ erg/sec).

ea3 Venus The energy absorption of Venus in ergs per gram per second (approximately $1.134E^{12}$ erg/sec).

EeS Earth's orbital kinetic energy (approximately $2.653E^{40}$ erg/sec).

electromagnetic energy (EME) The propagation of energy through the media Universe of Energy causing electric and magnetic field changes. Also called radiation and/or light. This radiation frequency spectrum varies from below, to far above the ability to be measured, but also includes the measurable spectrum of seconds per cycle to greater than $5.0E^{20}$ Hz (including visible light).

electron A theoretical particle with a "rest" mass of approximately $9.1094E^{-28}$ gram, and a negative charge of about $1.60219E^{-19}$ coulomb.

electron, free An electron that is not associated with or attached to an atom.

elliptical orbits Apparent orbits slightly out of round. Orbits are not round, but generally move out of round to become somewhat "oval" or parabolic (such as comet's orbits) due to anomalies of the UOE. Not true orbital paths.

emissions, human body Natural discards, secretions and emissions from the human body.

energy A power to create force, work, motion and change in mass characteristics and states. Philosophically, energy is also a mathematical concept derived from the indicators of mass characteristics.

energy density See density, energy.

energy pressure The force exerted on mass from energy as it reacts when near mass or entering mass.

energy to plant Photosynthesis, and other processes of plants that convert energy into mass.

energy transfer The conversion of one type energy to another, or from energy to mass, or mass to mass, or mass to energy.

energy wavelengths The distance between peak to peak or trough to trough of EME. The reciprocal of the frequency, then multiplied by c and a factor depending on the units required (such as meters, centimeters, etc.).

energy, electron orbital The kinetic energy required to maintain an electron orbit (as determined by the NRT formula).

entirety The whole (total) of the universe.

entity Something existing which is independent and unrelated to anything else.

enumeration Counting off, listing or cataloging.

equinox, autumnal See autumnal equinox.

equinox, vernal See vernal equinox.

Esur Earth's surface area (approximately $5.112E^{18}$ cm^2).

Evol Earth's volume (approximately $1.084E^{27}$ cm^3).

expanding universe See big bang.

expansion The decreased pressure on the trailing side of an object moving through the UOE as compared to the leading side (see compression).

FE The UOE energy entering earth.

fE The UOE energy that has entered, penetrated and left earth, minus the energy absorbed by the earth.

fission Splitting of an atomic nucleus into two or more parts.

FM The UOE energy entering the moon.

fM The energy that has entered, penetrated and left the moon, minus the energy absorbed by the moon.

fragulators Slang for competent Mathematicians, Physicists, Astrophysicists, or Scientists.

frequency The rate of oscillations in cycles. May be measured by cycles per second, Hertz (Hz, MHz, etc.), or seconds per cycle.

fusion, nuclear The combining of atomic nuclei (normally fusion is only created through applications of very high temperatures, such as in the millions of degrees Fahrenheit).

fusion, zero-point (ZPF) The UOE theoretical proposition that fusion may be obtained by lowering the energy of atoms through the reduction of heat until these atoms may be merged, then returning these merged atoms to normal temperature, thus causing fusion.

gamma energy shielding The absorption of gamma radiation by shielding material, such as concrete, water or lead.

generation, mass The creation of mass (LRE) during the velocity cycle of energy.

googolplex A number raised to the power googol ($X.0E^{100}$).

gravity The Universe of Energy. Includes all EME.

hadron A sub atomic particle with a strong interaction. Hadron class includes neutrons, protons, and pions.

half life The rate at which an object reduces itself, either physically or by radioactive decay. The reduction to one half life would leave exactly one half of the original amount.

heavy atomic mass Elements with a nucleus mass (A) of about 101 or greater.

Hertz Cycles per second.

high and low pressure In meteorology, the atmosphere consists of areas somewhat divided into different pressures. These pressures effect (cause) weather
and wind.

ine An atom's innermost orbital electron.

inner two thirds of earth Measured by volume (approximately $7.0E^{26}cm^3$).

inter-energy action Different types of energies affecting the state, velocity, or Tv
of other energies.

inverse square energy The energy to maintain a particle's satellite orbit, measured
or calculated at this particle's center.

inverse square law A formula for determining the changes (variations) of energy
amplitudes due to changes in distances between objects.

$$I_1 \times (D_1)^2 = I_2 \times (D_2)^2$$

where:

I_1 is original energy intensity at object 1

D_1 is original distance between objects 1 and 2

I_2 is second energy intensity at object 2

D_2 is second distance between objects 1 and 2

Note: I_1 or I_2 may have substitute designations such as E_1, E_2 or others used
by various disciplines of mathematics and physics.

Kenites Descendants of Cain.

Latent Reactive Energy (LRE) The energy construct of particles, referred to as
mass.

leading side The side of earth or (other parcel) facing toward the direction of
orbit. The earth's leading and trailing sides are constantly changing as the
earth rotates. See trailing side.

Lecton The hypothetical planet the size of an electron, used as a comparison for
the relationship of time frames and velocities.

life span The length of existence of a particle, a specific energy, or a living
creature.

light atomic mass Elements with a nucleus mass (A) of about 100 or less.

light meter An instrument used for measuring light intensity, normally for pho-
tography enhancement.

LRE See Latent Reactive Energy.

mass equivalent Mass that is not real, but only what could be as calculated by the
"$E = mc^2$" formula.

mass increase An increase of mass due to the velocity of mass, calculated by the
formula:

$$m = mo/ \sqrt{1 - (v/c)^2}$$

where: m is total mass, including the mass increase. mo is original mass
   v is velocity of mass c is velocity of light

mass increase (UOE) Mass increase at the rate of πE^{-20} cm³/sec, due to the absorption of Universe of Energy.

mass number The total mass (or weight) of an atom's nucleus, including protons and neutrons.

mass, permanent Mass maintaining a velocity that is constantly below the velocity of light. See latent reactive energy (LRE).

mass, photon Approximately $3.68625E^{-48}$ gram.

mathematician A person learned or skilled in mathematics (very few indeed).

max ever fast The hypothesis or assumption that velocities greater than light velocity are not achievable.

max faster The hypothesis or assumption that velocities greater than light velocities can be achieved.

mechanics, quantum Mathematical applications of the quantum theory. See quantum theory.

mE Earth's mass (approximately $5.9742E^{27}$ gram).

MED mass/energy density of the UOE ($6.454E^{12}$ gm/cm³).

media (medium) Either mass or energy intervening in the transmission or propagation of energy/mass.

MeE The energy required for the moon's orbit of earth.

meson A subclass of the hadrons. A meson may possess a charge with the magnitude of an electron, this charge may be positive, negative, or neutral.

meteor A small planet in space (normally associated with entry into earth's atmosphere showing a luminous trail).

meteorite A meteor that has reached the earth's surface.

meteorologist A "prophesier" or "soothsayer" of weather, performed scientifically.

meteorology The hypothetical science of predicting the weather, and dealing with the phenomena of earth's atmosphere.

MeV One million electron volts.

microgravity The gravity exerted on an object which is in an area that is neutral gravity (in- between two particles) or is under the inertia presented by a balanced orbit.

micronova Fissioning of atomic nucleus (two or more).

micronuclear explosion (MNX) Nomenclature representing all type novas.

microsecond One millionth of a second ($1.0E^{-6}$ sec).

min$_e$ velocity The calculated velocity of an atom's slowest outer orbiting electron.

mininova Fusion of atomic nuclei, one or many.

m$_\lambda$ Amount of transverted mass in gram per wavelength (7.3725E^{-48} g/l).

m$_\lambda$' Gram of mass per Transversion, two per wavelength (3.68625E^{-48} g/Tv).

L$_m$ Length in centimeter of transverted mass per wavelength.

m$_s$ Mass per second in gram (7.3725E^{-48} g x Hertz).

Mm The earth's moon's mass (approximately 7.3483E^{25} g)

MNX See micronuclear explosions.

molecular strong surface The molecular compression caused by the Universe of Energy. Tension can be measured in erg/cm^2, and the total force can be calculated by the surface area (cm^2) multiplied by the tension. Also referred to as surface tension.

MvE Moon's orbital velocity around earth (approximately 1.02321E^5 cm/sec or 2,289 mph, relatively).

np The average dimensions of the combination of a proton and neutron (mass, volume, etc.). A hypothetical number to reduce the time and improve ease of calculations that are based on neutrons, protons, and nuclei.

natural radiation The radiation emitted by the natural radioactive elements (isotopes) in the earth and all its inhabitants, and the radiation from the sun, stars, and all of the universe.

neutrino A neutrally charged particle exhibiting a quantum spin of 1/2. Neutrinos are commonly caused by light variation cycles or higher energies cycling in and out of mass **velocity** See Transversion.

neutron A particle having a mass number of one and a neutral charge. Commonly thought of as a proton and electron combined.

non-relativistic translational formula (NRT) A formula for calculating mass, energy, or velocity relating to particles with velocities less than 10% of light velocity, and moving in a linear direction. $E = 1/2mv^2$ (kinetic energy).

nova An incomplete or partial micronuclear explosion of a star or atomic nucleus.

nuclear force The repulsion of the constituents of a nucleus, periodically absorbing enough energy to partially "break through" the UOE binding force sending energy and/or particles out of the nucleus.

nucleus The center of an atom. The UOE concept considers the nucleus a plasma type particle exhibiting the combined reactions of protons and neutrons.

orbit anomalies See anomalies, orbit.

oute- The outer most orbital electron of an atom.

parallel UOE force The energy traveling parallel to the orbital path of a particle, either opposite or same direction.

parcel A particle or object in motion.

partial nova A sun flare (as an example).

particle Any object of any size that is considered mass.

particle expansion The "collection" (absorption) of energy by a particle which is transformed into mass. See mass increase.

particle origination A particle is created by energy that has been reduced to a velocity less than light velocity. The type, size, state, etc. of a particle is determined by the amount and type of energy that is involved in this velocity reduction.

particle, subatomic Particles that make up atomic nuclei, or is smaller than an atom.

peak The highest point in a wave crest, as opposed to the trough, which is the lowest point in a wave.

permanent mass See mass, permanent.

perpendicular entry An object or energy entering a planet (or particle) with a direct path line toward the center of this planet.

photon A particle equal to 1/2 Planck's constant divided by c^2, or about 3.68625E^{-48} gram. A photon is created by one single transversion of electromagnetic energy. The number of photons in a particle determines the particle's frequency (Hz) and the mass of the particle.

photon mass See mass, photon.

photosynthesis The conversion process green plants containing chlorophyll utilize to transform light energy to mass (chemical energy) and convert inorganic compounds to organic compounds, releasing oxygen.

piconova Unstable nuclei (radioactive).

picosecond One trillionth of a second ($1.0E^{-12}$ sec).

PIG formula A formula for calculating the probability of growth of particles:

$$tsec = \sqrt{mc/d}$$

where:

    t is time in seconds
    m is mass in gram
    c is velocity of light
    d is particle density

pion A subatomic particle in the meson family with a mass of about 264 times the mass of an electron, but with zero charge and a lifespan of about $9.0E^{-17}$ sec.

Pk See peak.

Planck's constant A constant representing the relationship of the quantum energy to frequency, expressed by: $E = hv$ Where: E is the energy of the quantum, v is the frequency, and h is Planck's constant of about $6.6260755E^{-27}$ erg sec.

plasma Mass at extreme temperatures from hundreds of thousands of degrees Fahrenheit to millions of degrees Fahrenheit, creating ionized gas, ions, free electrons, neutral particles, electron stripped nuclei, and great distances between molecules.

prefix A short nomenclature that indicates multiples or submultiples of a unit (such as micro, nano, mega, kilo, etc.).

probability in growth See PIG formula.

proportional A mathematical constant ratio. Variations of the same rate, either directly or inversely.

proton A positively charged particle with a mass about 1836 times the mass of the electron ($\sim 1.67252E^{-24}$ gram).

pure matter Hypothetical existence of 100% the energy within a specific area converted to mass.

quadrillion The cardinal number represented by 1 and followed by 15 zeros ($1.0E^{15}$ or 1,000,000,000,000,000)

quadrillionth The ordinal number represented by a decimal followed by 14 zeros and a 1 ($1.0E^{-15}$ or 0.000,000,000,000,001)

quantum theory The theory that electromagnetic radiation is constructed of photons, interacting in quanta. The amplitude, or magnitude of quanta may be calculated by multiplying Max Planck's constant (h) by the frequency (v or f) of the radiation.

quantum mechanics See mechanics, quantum.

quasi-orbits Particle orbits that are cycloid or sinusoidal rather than circular or elliptical due to the combined motion of the system. Earth's orbit is "quasi" due to the uniform and relative direction of motion of our solar system.

quintillion The cardinal number represented by 1 and followed by 18 zeros ($1.0E^{18}$ or 1,000,000,000,000,000,000).

quintillionth The ordinal number represented by a decimal followed by 17 zeros and a 1 ($1.0E^{-18}$ or 0.000,000,000,000,000,001).

radio frequencies Electromagnetic radiation with frequencies ranging from about $1.0E^4$ Hz to about $1.0E^{12}$ Hz.

radioactive elements "Unstable" elements (such as uranium, plutonium, radon, etc.) that emit energies in forms of electromagnetic radiation and particles. See nuclear force.

radius, apparent The surface area of an object as presented to an observer.

radius, electron (re-) The distance from the center of the electron to the outer edge (approximately $2.81777E^{-13}$ cm).

random "Haphazard" action or behavior that cannot be calculated, predicted, or estimated. Action beyond our knowledge of prediction.

re-emission The energy returned (repelled) by an object that cannot transform the arriving energy into heat, and dissipate this heat as rapidly as this energy is penetrating this object.

reduced force of energy Energy that continues on after part of it has been transferred to other energy or mass.

reflection See re-emission.

rest mass Also called rest energy. The mass of a body as observed while it is not in motion, relative to the reference system. There is no true rest mass, for all bodies are in motion, therefore energy is associated with mass at all times. If mass had no energy if would have no motion. $E = mc^2$ is generally an accepted formula for rest mass as compared to the NRT formula ($E = 1/2m \times v^2$) which is used for mass in motion.

reverse force A theoretical force that tends to slow down heavier elements more than light elements in a free fall.

RMS Root mean square. A trigonometrically calculated, or measured amplitude of a sine wave. RMS is also the value used to calculate electrical power in alternating current.

rotation of storms The resultant direction of storms' spin started by the UOE and the rotation of the earth. In the northern hemisphere storms normally rotate counterclockwise, and rotate clockwise in the southern hemisphere.

round off The shortening of large or long numbers, such as "8.2049857" to "8.205."

rpm Revolutions per minute.

SaE Sun's absorbed energy (approximately $1.22698E^{45}$ erg/sec).

Sd Sun's density (approximately 1.412 g/cm^3).

Sea1 Sun's absorbed energy per square centimeter.

Sea2 Sun's absorbed energy per cubic centimeter.

Sea3 Sun's absorbed energy per gram—approximately $2.98E^{12}$ erg/sec/g.

Sea4 Sun's total absorbed energy (approximately $5.93366E^{45}$ erg/sec).

shadow The effect of a particle blocking (shielding) a small section of the UOE from another particle.

shell The theoretical conception that an atom would resemble a solid ball due to the orbital electrons moving at velocities that would show them as only a blur, or a solid (see atomic shells).

shielding Material Material used to absorb radiation.

shielding area The effective area of an object for absorption of radiation.

sin Short for sine. In a Cartesian coordinate system the end point ordinate of an arc of the center of origin. Also, calculated in a triangle as the opposite side divided by the hypotenuse side.

sine wave variation The action occurring as oscillations in the sine wave function.

sinusoidal gravity The variations in gravity due to the rotation of the earth.

size/density variation References the expansion or contraction of a particle in relation to its density.

size state The time reference of a particle in relation to its size.

Sm Sun's mass (approximately $1.991E^{33}$ g).

sphereate The action of an object becoming shaped and rounded into a sphere shape.

Spr1 An imaginary sphere with its radius extending from the earth's center to the far side of the sun (approximate radius of $1.509E^{13}$ cm).

Spr1vol Spr1 volume (approximately $1.439E^{40}$ cm³).

Ssur sun's surface area (approximately $6.074E^{22}$ cm²).

subatomic particles The constituents of an atom, or smaller particles.

sub light velocity Velocities less than $2.99792458E^{10}$ cm/sec.

sun density variations The reduced densities of the sun as is measured, or calculated, from the center to the outer edges. The density at the center is much greater than near the surface.

sun's effective density The density as calculated from the sun's absorbed energy (approximately 3.697 g/cm³).

sun flare A partial nova.

sun mass See Sm.

sun standard density See Sd.

supercalculistic A mathematician or physicist.

supercomputerlators High tech, high power computers.

supernova A complete star explosion.

superstate A state in which a particle (usually an electron) appears to be in two or more different places at the same time.

surface tension See molecular strong surface.

tangent The right triangle acute angle function (angle = ratio of the opposite side to the adjacent side).

tectonic plates The structure of earth's crust and its deformities.

trailing side The side of earth or other parcels facing opposite their direction of orbit. The earth's leading and trailing sides are constantly changing as the earth rotates (see leading side).

transfer rate The time elapsed for energy/mass transforms to occur.

transform The change from energy to mass, mass to energy, energy to energy, or mass to mass.

Transversion (Tv) The transformation (conversion) of energy to mass and mass to energy due to the sinusoidal velocity variation of EME. See Tv.

T_m The duration of mass existence during a cycle of one energy wavelength.

T_m' The duration of mass existence during one transversion.

Tu A simulated particle used for an explanation.

Tv Transversion of mass to energy or energy to mass due to the sine function of EME velocity, occurring at inflections.

U238 A uranium atom (element), naturally occurring and most common, with a half life of about $4.51E^9$ yr.

uniforce A force that connects two or more forces, the Universe of Energy.

UOE The Universe of Energy, which encompasses the entire universe as energies of all types and velocities. The basis and media of all energy and matter (mass) creation, motion, and propagation. The UOE is also nuclear binding energy, week force, and gravity.

variations of UOE The energy amplitude variation due to absorption of energy by particles such as planets and stars. The majority of these variations are occurring at frequencies much greater than visible light velocity.

variations of gravity Earth's gravity variations are due to the various densities throughout the earth, sinusoidal variations and the variations in the UOE.

variations, light Light variations are due to light velocity frequency.

variation, weight A sinusoidal weight change during the rotation of the earth.

velocity, maximum The velocity of an inner orbiting electron. For the $92U^{238}$ atom this velocity is about $1.44E^8$ cm/sec.

velocity, minimum The velocity of an outer orbiting electron. For the $1H^1$ atom this velocity is about $1.384E^4$ cm/sec.

vernal equinox A point where the sun passes the celestial equator as its apparent motion is from south to north, starting the beginning of spring (around March 20, 21 or 22).

WAG system Wild Ass Guess. The logical method that is feasible for utilization by anyone from the uneducated layman to the most trained professor.

wavelength See energy wavelength.

wave packet A particle and its associated electromagnetic propagation. The ratio of particle (mass) to wave is determined mostly by the particle's velocity.

waveform A wave is a disturbance of a medium that will propagate (transmit) this disturbance. A waveform is the shape of this disturbance as plotted in time. There are many standard waveforms used in the field of electronics, such as a square wave, a sawtooth wave, and the common sine wave. There are many other natural waveforms.

zero degree, absolute The theoretical temperature at which molecular motion and particle repulsion ceases (-273.15 °C, or -459.67 °F).

zero gravity See microgravity.

zero-point fusion See fusion, zero-point.

zillion An excessively large, but indefinite number.

Index

Index of Illustrations

978-0-595-48881-0
0-595-48881-1

www.ingramcontent.com/pod-product-compliance
Lightning Source LLC
Chambersburg PA
CBHW030918180526
45163CB00002B/377